Analog Filter Design

<u>Electrical and Electronic Engineering Design Series</u>
Books written by Nicholas L. Pappas, Ph.D.

Electric Circuits – Analysis and Design

Electronic Circuit Design – with Bipolar and MOS Transistors

CMOS Circuit Design – Analog, Digital, IC Layout

Digital Design – Logic, Memory, Computers

Analog Filter Design

<u>Mathematics</u>
Books written by Nicholas L. Pappas, Ph.D.

Arithmetic – Integers, Fractions, Decimals

Analog Filter Design

Nicholas L. Pappas, Ph.D.

ISBN-13: 9781502554086   ISBN-10: 1502554089

A Message about this Text: The subject is essentially endless. The purpose here is to say enough about the subject so that you, the reader, have a running start when you apply this knowledge to your work.

Electronic circuit analysis and algebraic skills are prerequisites.

We believe important benefits accrue by doing the problems carefully, by reconstructing the Spice programs and running them to reproduce the text figures, by deriving the text equations, and by doing the experiments. These efforts provide *startup* work experience.

Once you have some work experience we are confident that you will be able to expand your know how with reasonable effort.

A Message from the Author: I have worked continuously in the electronics industry since 1950 except for 11 semesters teaching at San Jose State University (Professor and Chair Computer Engineering 1988-1993). There I discovered my talent for teaching such as it may be. After War2 I attended Lehigh University, and then transferred to Stanford where I earned the MS degree and, while working at HP in the early 1950's, the Ph.D. EE degree. (Somehow I did not get the word and formally apply for the BS degree.) Hardware design has been my principal activity. I learned enough about assembly language, Forth, C and C++ to design the software I needed for my projects. My current activity is designing integrated circuits.

Preface

An analog filter design process is presented here for (1) the Bell Telephone Laboratories constant k, and m derived ladder filters. (An n derived format is available for experimentation.), and (2) the modern Butterworth, Bessel, Chebyshev, and Inverse Chebyshev approximations of ideal filter transfer functions and their synthesis as circuits. *The designs produce filters one can build and use.* Spice programs verify performance.

The text starts with a presentation of the properties of four terminal two port networks. Voltage and current equations for network parameters z, y, h, g, abcd, and ABCD are developed. A parameter conversion table facilitates converting from one parameter type to another. Transfer functions and Thevenin equivalent circuits are produced. Voltage and current equations for various series, parallel, cascade network inter-connections are also produced. The equations and tables provide significant support for the filter design processes.

The equations of the Bell Telephone Laboratories LC ladder filters are developed in a straightforward manner. The underlying idea is that of *image impedance z_I,* which allows for cascading of individually designed filter sections while maintaining their transfer functions as designed. Cascading allows a designer to assemble filters with a wide variety of transfer functions. The ladder filter sections presented are constant k and m derived. Spice programs plot filter transfer functions. The lattice filter structure is *not* discussed.

The design of modern LC analog filters starts by showing how filters are specified. The specifications as written are not physically realizable by a circuit with real components. Approximations to the specifications such as Butterworth, Bessel, Chebyshev, and Inverse Chebyshev are implemented. The approximations are physically realizable rational functions of complex frequency variable p. In general, approximations produce a ratio of two polynomials in p, which we write as T(p) where T(p) is the output/input transfer function of a low pass filter. The next step is synthesis of the transfer function T(p) into an LC filter circuit. There are several synthesis methods. We present the Z transfer impedance method, and the Darlington insertion loss method. Spice programs plot filter transfer functions to verify the designs.

Analog Filter Design

Equations show how low pass filters can be converted into high pass, band pass, and band reject filters.

Transient response shows how various types of filters modify a step function.

We show how to write Spice programs that document filter performance via alternating voltage and current (AC) analysis and TRAN transient response.

We include useful *experiments* that give you real world experience. We consider the experiments to be significant learning activities.

The experiments include elementary RLC filters, Bell Telephone Laboratories filters, active filters using op amps, and filters derived via approximations.

Chapter abstracts are next followed by a historical note.

> Our blog *npappasee.blogspot.com* may offer you additional information. Take a look.

> We would appreciate receiving your comments and views on this text at *npappasz@yahoo.com*.

1 Four Terminal Two Port Networks This topic facilitates the design of filters, and other complex circuits.

Four terminal network theory provides means to examine the behavior of a circuit while knowing essentially nothing about the details of the circuit. By focusing on the input and output terminals of a circuit (a black box if you will) the representation and, or specification of the circuit is straightforward. Furthermore, once individual four terminal networks are synthesized to have desired properties, they are readily assembled into series, parallel, and cascaded combinations.

A four terminal network is a circuit accessible via two pairs of terminals. The terminals' two voltages and two currents allow the networks' external behavior to be represented in six ways by six pairs of equations relating port voltages and currents. The pairs of equations define sets of parameters such as the system of z's, which are open circuit driving point and transfer impedances. The z, y, h, g, abcd, ABCD parameters relationships are developed. Most of the information is for reference as needed.

The process designing analog filters makes extensive use of the four terminal network properties and equations.

2 Ladder Filter Design The Bell Telephone Laboratories ladder filter sections constant k, and m derived equations are developed in a straight-forward manner. The different types of sections have the same *image impedance* z_I by design, which allows for cascading of individually designed filter sections while maintaining their transfer functions as designed.

The general *propagation function* for sections is derived. This ratio of output voltage to input voltage is $e^{-\gamma}$. The T and π *image impedance* equations are developed.

Cascading with constant k, and m derived sections allows a designer to assemble filters with a wide variety of transfer functions.

Low Pass, high pass, and band pass design tables facilitate the design process.

3 Modern Filter Design Practical design processes are presented as straightforward procedures. We start by showing how filter frequency responses are specified. The first problem to solve is finding a physically realizable approximation to the specified filter frequency response. We show how to implement the Butterworth, Bessel, Chebyshev, and Inverse Chebyshev approximations, which are physically realizable rational functions of p. Approximations produce a ratio of two polynomials in p $N_n(p)/D_n(p)$, which we write as $T_n(p)$ where $T_n(p)$ is the output/input transfer function of a low pass filter. Integer n is the degree of the approximation polynomial of the filter transfer function.

Every filter has a transition from the pass band to the stop band. The slope of the transition is essentially $-20n$ dB per decade.

4 Modern Filter Synthesis The second problem to solve is implementing transfer function $T_n(p)$ as a circuit. There are several possibilities. We present the transfer Z impedance synthesis method, and the Darlington insertion loss synthesis method. Both methods produce ladder networks.

Frequency transformations convert low pass filters into high pass, band pass, and band reject filters. Spice programs document filter performance.

5 Filter Transient Response Filter Transient response shows how various types of filters modify a step function.

6 How to write Spice Programs Spice is used to plot performance of AC frequency response and TRAN transient response of filter circuits that are analyzed and designed in the text. We recognize that you do not have to crunch numbers today, because Spice is the modern way to crunch numbers.

Eight Experiments
Experiment 1 Elementary RLC Filters
Experiment 2 Design of Ladder Filters
Experiment 3 Design an Op Amp Low Pass Filter
Experiment 4 Design an Op Amp High Pass Filter
Experiment 5 Design an Op Amp Band Pass Filter
Experiment 6 Design a Butterworth Filter
Experiment 7 Design a Bessel Filter
Experiment 8 Design a Chebyshev Filter

Historical Note

Filters evolving from Transmission Lines We start here with the ideas behind the *electric wave filters* of Campbell[1], Zobel[2] and Wagner[3]. Their filters are derived from lumped approximations to transmission lines.

Consider the driving point impedance z_{in} of the infinite ladder network:

Figure 001. Ladder Network representing a transmission line

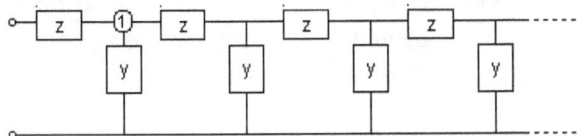

If the ladder network input impedance is z_{in}, then the impedance looking to the right of node 1 (Figure 001) also equals z_{in}, because this network has an infinite number of sections. Therefore the infinite network can be replaced by a finite network (Figure 002).

Figure 002

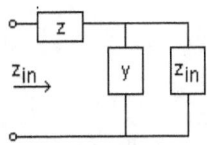

$$(1) \quad z_{in} = z + \cfrac{1}{y + \cfrac{1}{z_{in}}} = z + \frac{z_{in}}{1 + yz_{in}} = \frac{z(1 + yz_{in}) + z_{in}}{1 + yz_{in}} = \frac{z + (1 + yz)z_{in}}{1 + yz_{in}}$$

$$z_{in}(1 + yz_{in}) = z + (1 + yz)z_{in} \;\rightarrow\; yz_{in}^2 + z_{in} = z + z_{in} + yzz_{in} \;\rightarrow\; z_{in}^2 - zz_{in} - \frac{z}{y} = 0$$

$$z_{in} = \frac{z}{2} \pm \sqrt{\frac{z^2}{4} + \frac{z}{y}} = \frac{z}{2} \pm \frac{z}{2}\sqrt{1 + \frac{4}{yz}}$$

$$(2a) \quad \text{let } z = pL = j\omega L \qquad y = j\omega C \qquad zy = -\omega^2 LC$$

$$(2b) \quad z_{in} = \frac{j\omega L}{2}\left(1 + \sqrt{1 - \frac{4}{\omega^2 LC}}\right)$$

[1] G. A. Campbell, *Physical Theory of the Electric Wave-Filter*, *Bell System Technical Journal*, vol. 1, no.2, pp. 1-32, Nov. 1922.
[2] Zobel, O. J.,*Theory and Design of Uniform and Composite Electric Wave Filters*, Bell Systems Technical Journal, Vol. 2 (1923), pp. 1–46.
[3] Wagner, *Spulen- und Kondensatorleitungen*, Archiv für Electrotechnik, vol. 8, July 1919.

Analog Filter Design

At very low frequencies the term under the radical in equation 2b is negative so that the square root is imaginary and greater than 1. Consequently z_{in} is real and equal to a constant k referred to as *constant k* by Zobel.

$$(3) \quad z_{in} = \frac{j\omega L}{2}\left(1+\sqrt{1-\frac{4}{\omega^2 LC}}\right) \approx \frac{j\omega L}{2}\left(\frac{2}{j\omega\sqrt{LC}}\right) = \sqrt{\frac{L}{C}} = k \quad when \quad \frac{4}{\omega^2 LC} \gg 1$$

A z_{in} that has a real component will pass power from input to output. However above the *cutoff* frequency ω_c the input impedance z_{in} becomes imaginary, and no power is transferred to the output.

The *pass band* is from 0 to ω_c, and the *attenuation band* is from ω_c to infinity. This is a *low pass* filter.

$$(4) \quad z_{in} = \frac{j\omega L}{2}\left(1+\sqrt{1-\frac{4}{\omega^2 LC}}\right) \approx \frac{j\omega L}{2}(1+1) \approx j\omega L \quad when \quad \frac{4}{\omega^2 LC} \ll 1$$

$$(5) \quad 1-\frac{4}{\omega_c^2 LC} = 0 \;\rightarrow\; 1 = \frac{4}{\omega_c^2 LC} \;\rightarrow\; \omega_c^2 = \frac{4}{LC} \;\rightarrow\; \omega_c = \frac{2}{\sqrt{LC}} = 2\omega_0$$

The analysis above assumed the filter includes an infinite number of zy sections, because the filter is terminated in z_{in}. On the other hand most filter designs include a finite number of sections, and they are terminated by a resistor R. Nevertheless the ideas of pass bands, attenuation bands, (image impedance) z_{in}, and cutoff frequency remain valid.

Modern Filters – Approximations and Synthesis Control over filter response is achieved by direct manipulation of transfer function pole and zero locations. Butterworth, Bessel, Chebyshev, and Inverse Chebyshev polynomial approximations to specified filter transfer functions are synthesized by straightforward processes. For example here is a low pass filter Butterworth polynomial approximation of degree n=3

$$(6) \quad T_{bu3}(p) = \frac{1}{p^3 + 2p^2 + 2p + 1}$$

Contents

Greek Alphabet

A	α	alpha	a[10]
B	β	beta	b
Γ	γ	gamma	g
Δ	δ	delta	d
E	ε	epsilon	e
Z	ζ	zeta	z
H	η	eta	h
Θ	θ	theta	q
I	ι	iota	i
K	κ	kappa	k
Λ	λ	lambda	l
M	μ	mu	m
N	ν	nu	n
Ξ	ξ	xsi	x
O	o	omicron	o
Π	π	pi	p
P	ρ	rho	r
Σ	σ	sigma	s
T	τ	tau	t
Y	υ	upsilon	u
Φ	ϕ	phi	f
X	χ	chi	c
Ψ	ψ	psi	y
Ω	ω	omega	w

---

[10] equivalent computer keyboard English letter keys

1 Four Terminal Two Port Networks

A four terminal network (Figure 101) is a *black box* with two terminals designated as inputs and two terminals designated as outputs. Each pair of terminals is also referred to as a *port*. The box contains a linear network with these properties.

Figure 101

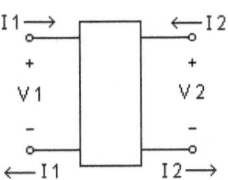

1. Initial stored energy is zero.
2. Current into a port terminal equals current out of the other port terminal.
3. All external connections are made to the port terminals.
4. There are no independent sources in the black box (dependent sources are allowed).

The internal circuit is not a concern when four terminal networks are studied, because the emphasis is on the network's voltage current vi constraints. The discussion is limited to two ports, even though a circuit can have many ports each with input impedance, and transfer impedance's relating to each of the other ports.

Four terminal network theory provides means to examine the external behavior of a circuit. By focusing on the input and output terminals of a circuit the representation and, or specification of the circuit is simplified. Furthermore, once individual four terminal networks are synthesized to have desired properties. They are readily assembled into series, parallel, and cascaded combinations.

Major results derived from four terminal two port network theory are filter characteristic impedance z_0 and propagation function e^y.

The four terminal network's two voltages and two currents allow the network to be represented in six ways by six pairs of equations ($4 \times 3 \times \frac{1}{2}$). The pairs of equations, in effect, define sets of parameters such as open circuit input and transfer impedances. We show how to set up the equations as well as how to find the equations for the various types of parameters.

Analog Filter Design

Relationships are derived between parameter sets, because they can expedite circuit analysis. E.g., the open circuit impedance's may be the easiest to calculate, but the hybrid parameters may be the most appropriate for understanding the circuit behavior.

The four terminal network is placed in a practical context: it is driven by a source with finite impedance z_S and terminated with load impedance z_L. Important characteristics such as input impedance, transmission from source to load, and Thevenin impedance and voltage are derived.

Practical examples illustrate how four terminal network theory is used.

Many nodes Consider a circuit with four nodes and two sources. Write the equations, while ignoring the port idea.

(1) $I_1 = y_{11}V_1 + y_{12}V_2 + y_{13}V_3 + y_{14}V_4$
$I_2 = y_{21}V_1 + y_{22}V_2 + y_{23}V_3 + y_{24}V_4$
$0 = y_{31}V_1 + y_{32}V_2 + y_{33}V_3 + y_{34}V_4$
$0 = y_{41}V_1 + y_{42}V_2 + y_{43}V_3 + y_{44}V_4$

We can solve these equations for each of the four voltages. However, we are interested in the solutions for V_1 and V_2, because we want to pursue the four terminal network idea. The following node circuit analysis shows that the coefficient of each current is an impedance (V=zI) derived from a ratio of two determinants of admittances. The Δ_{rc} determinant's subscripts r, c are row r and column c numbers. For example, Δ_{21} is the 3×3 minor formed from the 4×4 determinant Δ_y by striking out row 2 and column 1.

(2a) $V_1 = \dfrac{\Delta_{11}}{\Delta_y}I_1 + \dfrac{\Delta_{21}}{\Delta_y}I_2 \equiv z_{11}I_1 + z_{12}I_2$

(2b) $V_2 = \dfrac{\Delta_{12}}{\Delta_y}I_1 + \dfrac{\Delta_{22}}{\Delta_y}I_2 \equiv z_{21}I_1 + z_{22}I_2$

Note There is a reverse subscript relationship between subscripts of the z transfer impedances and the y determinants.

Even when there are many circuit nodes the solution for any voltage will have only two terms when there are only two sources driving the circuit.

1.1 Four Terminal Network Parameters

The four terminal network circuit is defined by the two voltages and two currents. If two of these four circuit variables are selected as independent variables then the other two variables are dependent variables. How many choices are there? The first independent variable can be selected in four ways; the second is selected from one of the three remaining variables. Thus twelve equations are formed as six pairs. The six sets of equations in Table 101 define six sets of parameters (z, y, A, a, h, and g). Any parameter set is derived by circuit analysis using the mesh or nodal method. We examine the z, y, A, and h parameters.

Table 101 Four terminal network Equations

$$\begin{vmatrix} V_1 \\ V_2 \end{vmatrix} = \begin{vmatrix} z_{11} & z_{12} \\ z_{21} & z_{22} \end{vmatrix} \times \begin{vmatrix} I_1 \\ I_2 \end{vmatrix} \qquad \begin{aligned} V_1 &= z_{11}I_1 + z_{12}I_2 \\ V_2 &= z_{21}I_1 + z_{22}I_2 \end{aligned}$$

$$\begin{vmatrix} I_1 \\ I_2 \end{vmatrix} = \begin{vmatrix} y_{11} & y_{12} \\ y_{21} & y_{22} \end{vmatrix} \times \begin{vmatrix} V_1 \\ V_2 \end{vmatrix} \qquad \begin{aligned} I_1 &= y_{11}V_1 + y_{12}V_2 \\ I_2 &= y_{21}V_1 + y_{22}V_2 \end{aligned}$$

$$\begin{vmatrix} V_1 \\ I_1 \end{vmatrix} = \begin{vmatrix} A & B \\ C & D \end{vmatrix} \times \begin{vmatrix} V_2 \\ -I_2 \end{vmatrix} \qquad \begin{aligned} V_1 &= AV_2 - BI_2 \\ I_1 &= CV_2 - DI_2 \end{aligned}$$

$$\begin{vmatrix} V_2 \\ I_2 \end{vmatrix} = \begin{vmatrix} a & b \\ c & d \end{vmatrix} \times \begin{vmatrix} V_1 \\ -I_1 \end{vmatrix} \qquad \begin{aligned} V_2 &= aV_1 - bI_1 \\ I_2 &= cV_1 - dI_1 \end{aligned}$$

$$\begin{vmatrix} V_1 \\ I_2 \end{vmatrix} = \begin{vmatrix} h_{11} & h_{12} \\ h_{21} & h_{22} \end{vmatrix} \times \begin{vmatrix} I_1 \\ V_2 \end{vmatrix} \qquad \begin{aligned} V_1 &= h_{11}I_1 + h_{12}V_2 \\ I_2 &= h_{21}I_1 + h_{22}V_2 \end{aligned}$$

$$\begin{vmatrix} I_1 \\ V_2 \end{vmatrix} = \begin{vmatrix} g_{11} & g_{12} \\ g_{21} & g_{22} \end{vmatrix} \times \begin{vmatrix} V_1 \\ I_2 \end{vmatrix} \qquad \begin{aligned} I_1 &= g_{11}V_1 + g_{12}I_2 \\ V_2 &= g_{21}V_1 + g_{22}I_2 \end{aligned}$$

Problem 101 For the bridged-T in Example 101 Show that when $i_1=0$

$$\frac{i_a}{i_2} = \frac{r_0}{pL_1 + 2r_0} \qquad where \quad v_2 - v_3 = i_a(pL_1 + r_0) = i_b r_0 \quad and \quad i_2 = i_a + i_b$$

Problem 102 Show that when $i_1=0$

$$z_{12} = \frac{v_1}{i_2} = \frac{i_a r_0}{i_2} + \frac{1}{i_2}\frac{i_2}{pC_1} = \frac{r_0}{pL_1 + 2r_0}r_0 + \frac{1}{pC_1}$$

1.1.1 Z Parameters

When currents I_1 and I_2 drive a circuit, Kirchhoff's node method results in a solution for voltages.

(3a)　$V_1 = z_{11}I_1 + z_{12}I_2$

(3b)　$V_2 = z_{21}I_1 + z_{22}I_2$

Observe that if $I_2 = 0$ then $z_{11} = V_1/I_1$. The z_{ij} are referred to as *open-circuit parameters* because they are found when there is an open circuit at an external port.

(4a)　$V_1 = z_{11}I_1 + z_{12} \cdot 0 \ \Rightarrow \ z_{11} = \left.\dfrac{V_1}{I_1}\right|_{I_2=0}$

(4b)　$V_2 = z_{21}I_1 + z_{22} \cdot 0 \ \Rightarrow \ z_{21} = \left.\dfrac{V_2}{I_1}\right|_{I_2=0}$

(4c)　$V_1 = z_{11} \cdot 0 + z_{12}I_2 \ \Rightarrow \ z_{12} = \left.\dfrac{V_1}{I_2}\right|_{I_1=0}$

(4d)　$V_2 = z_{21} \cdot 0 + z_{22}I_2 \ \Rightarrow \ z_{22} = \left.\dfrac{V_2}{I_2}\right|_{I_1=0}$

Example　101　Z　Parameters Find the z parameters of the *symmetrical bridged-T* circuit.

$V_1 = z_{11}I_1 + z_{12}I_2$

$V_2 = z_{21}I_1 + z_{22}I_2$

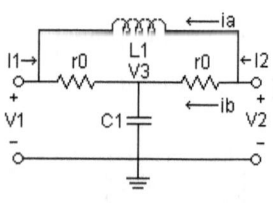

z_{11}　　With port 2 open circuited $I_2 = 0$,
　　　　port 1 sees r_0 in parallel with pL_1+r_0 to which is added $1/pC_1$.

z_{12}　　With port 1 open circuited $I_1 = 0$ and I_2 is the source, the port 1 voltage is
　　　　the drop across r_0 and $1/pC_1$.

z_{21}　　Equals z_{12}.

z_{22}　　Equals z_{11}.

$z_{11} = \left.\dfrac{V_1}{I_1}\right|_{I_2=0} = \dfrac{r_0(pL_1 + r_0)}{pL_1 + 2r_0} + \dfrac{1}{pC_1}$　　$z_{21} = \left.\dfrac{V_2}{I_1}\right|_{I_2=0} = z_{12}$

$z_{12} = \left.\dfrac{V_1}{I_2}\right|_{I_1=0} = \dfrac{r_0(r_0)}{pL_1 + 2r_0} + \dfrac{1}{pC_1}$　　$z_{22} = \left.\dfrac{V_2}{I_2}\right|_{I_1=0} = z_{11}$

1.1.2 Y Parameters

The solution to mesh equations driven by two voltage sources has the form

(5a) $I_1 = y_{11}V_1 + y_{12}V_2$

(5b) $I_2 = y_{21}V_1 + y_{22}V_2$

The y_{jk} admittances are referred to as *short circuit parameters*, because port terminals are shorted to make port voltages equal to zero.

(6a) $I_1 = y_{11}V_1 + y_{12} \cdot 0 \quad \Rightarrow \quad y_{11} = \dfrac{I_1}{V_1}\bigg|_{V_2=0}$

(6b) $I_2 = y_{21}V_1 + y_{22} \cdot 0 \quad \Rightarrow \quad y_{21} = \dfrac{I_2}{V_1}\bigg|_{V_2=0}$

(6c) $I_1 = y_{11} \cdot 0 + y_{12}V_2 \quad \Rightarrow \quad y_{12} = \dfrac{I_1}{V_2}\bigg|_{V_1=0}$

(6d) $I_2 = y_{21} \cdot 0 + y_{22}V_2 \quad \Rightarrow \quad y_{22} = \dfrac{I_2}{V_2}\bigg|_{V_1=0}$

Example 102 Y Parameters

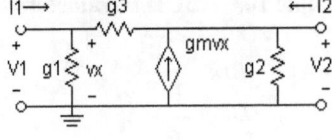

Find the y parameters of the Pi circuit with a dependent source.

$I_1 = y_{11}V_1 + y_{12}V_2$

$I_2 = y_{21}V_1 + y_{22}V_2$

y_{11} With port 2 short circuited so that $V_2 = 0$ port 1 sees g_1 and g_3 in parallel.

y_{12} With port 1 short circuited so that $V_1 = 0$ and V_2 is the source, the port 1 current $I_1 = -g_3V_2$.

y_{21} With port 2 short circuited so that $V_2 = 0$ and V_1 is the source, the port 2 current $-I_2 = g_3V_1 + g_mV_1$.

y_{22} With port 1 short circuited so that $V_1 = 0$ port 2 sees g_2 and g_3 in parallel. The dependent source current is zero because $v_x = 0$.

$y_{11} = \dfrac{I_1}{V_1}\bigg|_{V_2=0} = g_1 + g_3 \qquad\qquad y_{12} = \dfrac{I_1}{V_2}\bigg|_{V_1=0} = -g_3$

$y_{21} = \dfrac{I_2}{V_1}\bigg|_{V_2=0} = -(g_3 + g_m) \qquad y_{22} = \dfrac{I_2}{V_2}\bigg|_{V_1=0} = g_2 + g_3$

1.1.3 ABCD Parameters

The ABCD parameters are known as *transmission parameters*, because they relate input-port voltage and current to output-port voltage and current. The minus signs appear, because transmission implies $-I_2$ flows in the same direction as I_1: to the right.

(7a) $V_1 = AV_2 - BI_2$

(7b) $I_1 = CV_2 - DI_2$

(8a) $V_1 = A \cdot 0 - BI_2 \quad \Rightarrow \quad B = -\dfrac{V_1}{I_2}\bigg|_{V_2=0}$

(8b) $I_1 = C \cdot 0 - DI_2 \quad \Rightarrow \quad D = -\dfrac{I_1}{I_2}\bigg|_{V_2=0}$

(8c) $V_1 = AV_2 - B \cdot 0 \quad \Rightarrow \quad A = \dfrac{V_1}{V_2}\bigg|_{I_2=0}$

(8d) $I_1 = CV_2 - D \cdot 0 \quad \Rightarrow \quad C = \dfrac{I_1}{V_2}\bigg|_{I_2=0}$

Example 103 ABCD Parameters for a Practical Transformer

$V_1 - AV_2 - BI_2$

$I_1 = CV_2 - DI_2$

The ABCD parameters for a practical transformer are found by using the mesh method.

$V_1 = (R_1 + pL_1)I_1 + (pM)I_2$

$V_2 = (pM)I_1 + (R_2 + pL_2)I_2$

$I_2 = 0 \quad \Rightarrow \quad V_1 = (R_1 + pL_1)I_1 \quad and \quad V_2 = (pM)I_1$

$V_2 = 0 \quad \Rightarrow \quad V_1 = (R_1 + pL_1)I_1 + (pM)I_2 \quad and \quad 0 = (pM)I_1 + (R_2 + pL_2)I_2$

therefore

$A = \dfrac{V_1}{V_2}\bigg|_{I_2=0} = \dfrac{R_1 + pL_1}{pM} \qquad B = -\dfrac{V_1}{I_2}\bigg|_{V_2=0} = \dfrac{(R_1 + pL_1)(R_2 + pL_2) - (pM)^2}{pM}$

$C = \dfrac{I_1}{V_2}\bigg|_{I_2=0} = \dfrac{1}{pM} \qquad D = -\dfrac{I_1}{I_2}\bigg|_{V_2=0} = \dfrac{R_2 + pL_2}{pM}$

1.1.4 h Parameters

The h parameters are a mix of short circuit and open circuit parameters. They are suitable for specifying transistor models, for example. Many commercial transistor data sheets use h parameters.

(9a) $V_1 = h_{11}I_1 + h_{12}V_2$

(9b) $I_2 = h_{21}I_1 + h_{22}V_2$

h_{11} and h_{21} are short circuit parameters ($V_2 = 0$), and h_{12} and h_{22} are open circuit parameters ($I_1 = 0$).

(10a) $V_1 = h_{11}I_1 + h_{12} \cdot 0 \implies h_{11} = \left.\dfrac{V_1}{I_1}\right|_{V_2=0}$

(10b) $I_2 = h_{21}I_1 + h_{22} \cdot 0 \implies h_{21} = \left.\dfrac{I_2}{I_1}\right|_{V_2=0}$

(10c) $V_1 = h_{11} \cdot 0 + h_{12}V_2 \implies h_{12} = \left.\dfrac{V_1}{V_2}\right|_{I_1=0}$

(10d) $I_2 = h_{21} \cdot 0 + h_{22}V_2 \implies h_{22} = \left.\dfrac{I_2}{V_2}\right|_{I_1=0}$

Example 104   h parameters for a transistor model that includes a dependent source.

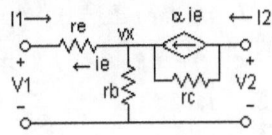

$V_1 = h_{11}I_1 + h_{12}V_2$

$I_2 = h_{21}I_1 + h_{22}V_2$

h_{11} Port 2 is short circuited so that $V_2 = 0$. To calculate h_{11} we need to do a nodal analysis to calculate V_x/V_1.

$-I_e = g_e(V_1 - V_x) = I_1$

$0 = -g_e V_1 + [g_e + g_b + g_c]V_x - g_c V_2 - \alpha I_e$

$0 = -g_e V_1 + [g_e + g_b + g_c]V_x + \alpha g_e(V_1 - V_x) - g_c V_2$

$0 = -g_e(1-\alpha)V_1 + [g_e(1-\alpha) + g_b + g_c]V_x - g_c V_2$

$\left.\dfrac{V_x}{V_1}\right|_{V_2=0} = \dfrac{g_e(1-\alpha)}{g_e(1-\alpha) + g_b + g_c}$

Continued on the next page

$$h_{11} = \frac{V_1}{I_1}\bigg|_{V_2=0} = \frac{V_1}{-I_e} = \frac{V_1}{g_e(V_1-V_x)} = r_e\frac{1}{1-\dfrac{V_x}{V_1}} = r_e\frac{1}{1-\dfrac{g_e(1-\alpha)}{g_e(1-\alpha)+g_b+g_c}}$$

$$h_{11} = \frac{V_1}{I_1}\bigg|_{V_2=0} = r_e\left[\frac{g_e(1-\alpha)+g_b+g_c}{g_b+g_c}\right] = r_e\left[1+\frac{g_e(1-\alpha)}{g_b+g_c}\right] = r_e+(1-\alpha)\frac{r_b r_c}{r_b+r_c}$$

h_{12} Port 1 is open circuited so that $I_1 = 0$. Since $I_1 = -I_e = 0$, V_1 equals the output of the r_b, r_c voltage divider driven by source V_2.

$$0 = -g_e V_1 + [g_e+g_b+g_c]V_x - g_c V_2 + \alpha I_1$$
$$\text{if } I_1 = 0, \text{ then } V_x = V_1 \text{ and so}$$
$$0 = -g_e V_1 + [g_e+g_b+g_c]V_1 - g_c V_2 = [g_b+g_c]V_1 - g_c V_2$$
$$h_{12} = \frac{V_1}{V_2}\bigg|_{I_2=0} = \frac{g_c}{g_b+g_c} = \frac{r_b}{r_b+r_c}$$

h_{21} Port 2 is short circuited so that $V_2=0$ and $I_1=-I_e$ is the source. The port 2 current $-I_2$ equals αI_1 plus r_c part of the remaining current $(1-\alpha)I_1$ that divides between r_c and r_b.

$$\text{if } I_1 = 0 \text{ then } I_e = 0, \ V_x = V_1, \text{ and so}$$
$$0 = -g_e V_1 + [g_e+g_b+g_c]V_1 - g_c V_2 = [g_b+g_c]V_1 - g_c V_2$$
$$h_{21} = \frac{I_1}{I_2}\bigg|_{V_2=0} = \frac{g_c}{g_b+g_c} = \frac{r_b}{r_b+r_c}$$

h_{22} Port 1 is open circuited so that $I_1=I_e=0$. Port 2 "sees" r_c in series with r_b because $\alpha I_e = 0$.

$$\text{if } I_1 = 0 \text{ then } I_e = 0, \ V_x = V_1, \ \Rightarrow \ 0 = -g_e V_x + [g_e+g_b+g_c]V_x - g_c V_2$$
$$0 = [g_b+g_c] \quad V_x - g_c V_2 (I_1 = 0) \Rightarrow \frac{V_x}{V_2} = \frac{g_c}{g_b+g_c}$$
$$I_2 = g_c[V_2-V_x] = g_c V_2\left[1-\frac{V_x}{V_2}\right] = g_c V_2\left[1-\frac{g_c}{g_b+g_c}\right] = g_c V_2\frac{g_b}{g_b+g_c} = \frac{V_2}{r_b+r_c}$$
$$h_{22} = \frac{I_2}{V_2}\bigg|_{I_2=0} = \frac{1}{r_b+r_c}$$

1.2 Four terminal network parameter relationships

Establishing relationships requires a straightforward use of algebra or matrix theory. For example converting from z to y parameters is achieved by writing the z equations and using Cramer's rule to solve for the currents, because the resulting equations have the form of y equations.

(11a) $V_1 = z_{11}I_1 + z_{12}I_2$

(11b) $V_2 = z_{21}I_1 + z_{22}I_2$

solve for the currents and equate to y parameters

Cramer found responses y_1, y_2 to forcing functions x_1, x_2

$\quad$ *If* $r_1 = a_{11}s_1 + a_{12}s_2$ *and* $r_2 = a_{21}s_1 + a_{22}s_2$

Then $\Delta = a_{11}a_{22} - a_{21}a_{12}$ *and*

$$s_1 = \frac{\begin{vmatrix} r_1 & a_{12} \\ r_2 & a_{22} \end{vmatrix}}{\begin{vmatrix} a_{11} & a_{12} \\ a_{21} & a_{22} \end{vmatrix}} = r_1 \frac{a_{22}}{\Delta} - r_2 \frac{a_{12}}{\Delta} \qquad s_2 = \frac{\begin{vmatrix} a_{11} & r_1 \\ a_{21} & r_2 \end{vmatrix}}{\begin{vmatrix} a_{11} & a_{12} \\ a_{21} & a_{22} \end{vmatrix}} = -r_1 \frac{a_{21}}{\Delta} + r_2 \frac{a_{11}}{\Delta}$$

(12a) $I_1 = +\dfrac{z_{22}}{\Delta_z}V_1 - \dfrac{z_{12}}{\Delta_z}V_2 = y_{11}V_1 + y_{12}V_2$

(12b) $I_2 = -\dfrac{z_{21}}{\Delta_z}V_1 + \dfrac{z_{11}}{\Delta_z}V_2 = y_{21}V_1 + y_{22}V_2$

(13c) $y_{11} = \dfrac{z_{22}}{\Delta_z}$ (13b) $y_{12} = -\dfrac{z_{12}}{\Delta_z}$ (13c) $y_{21} = -\dfrac{z_{21}}{\Delta_z}$ (13d) $y_{22} = \dfrac{z_{11}}{\Delta_z}$

Problem 103 Find the ABCD parameters for the circuit of Example 101.

Problem 104 Find the ABCD parameters for the circuit of Example 102.

Problem 105 Find the h parameters for the circuit of Example 102.

Problem 106 Convert the h parameters of Example 104 to z parameters.

Problem 107 Convert the Example 103 ABCD parameters to h parameters.

Analog Filter Design

Table 102 Four terminal network Conversion Table

	$|z|$		$|y|$		$|h|$		$|T|$	
$|z|$	z_{11}	z_{12}	$\dfrac{y_{22}}{\Delta_y}$	$-\dfrac{y_{12}}{\Delta_y}$	$\dfrac{\Delta_h}{h_{22}}$	$\dfrac{h_{12}}{h_{22}}$	$\dfrac{A}{C}$	$\dfrac{\Delta_T}{C}$
	z_{21}	z_{22}	$-\dfrac{y_{21}}{\Delta_y}$	$\dfrac{y_{11}}{\Delta_y}$	$-\dfrac{h_{21}}{h_{22}}$	$\dfrac{1}{h_{22}}$	$\dfrac{1}{C}$	$\dfrac{D}{C}$
$|y|$	$\dfrac{z_{22}}{\Delta_z}$	$-\dfrac{z_{12}}{\Delta_z}$	y_{11}	y_{12}	$\dfrac{1}{h_{11}}$	$-\dfrac{h_{12}}{h_{11}}$	$\dfrac{D}{B}$	$-\dfrac{\Delta_T}{B}$
	$-\dfrac{z_{21}}{\Delta_z}$	$\dfrac{z_{11}}{\Delta_z}$	y_{21}	y_{22}	$\dfrac{h_{21}}{h_{11}}$	$\dfrac{\Delta_h}{h_{11}}$	$-\dfrac{1}{B}$	$\dfrac{A}{B}$
$|h|$	$\dfrac{\Delta_z}{z_{22}}$	$\dfrac{z_{12}}{z_{22}}$	$\dfrac{1}{y_{11}}$	$-\dfrac{y_{12}}{y_{11}}$	h_{11}	h_{12}	$\dfrac{B}{D}$	$\dfrac{\Delta_T}{D}$
	$-\dfrac{z_{21}}{z_{22}}$	$\dfrac{1}{z_{22}}$	$\dfrac{y_{21}}{y_{11}}$	$\dfrac{\Delta_y}{y_{11}}$	h_{21}	h_{22}	$-\dfrac{1}{D}$	$\dfrac{C}{D}$
$|T|$	$\dfrac{z_{11}}{z_{21}}$	$\dfrac{\Delta_z}{z_{21}}$	$-\dfrac{y_{22}}{y_{21}}$	$-\dfrac{1}{y_{21}}$	$-\dfrac{\Delta_h}{h_{21}}$	$-\dfrac{h_{11}}{h_{21}}$	A	B
	$\dfrac{1}{z_{21}}$	$\dfrac{z_{22}}{z_{21}}$	$-\dfrac{\Delta_y}{y_{21}}$	$-\dfrac{y_{11}}{y_{21}}$	$-\dfrac{h_{22}}{h_{21}}$	$-\dfrac{1}{h_{21}}$	C	D

Symmetrical four terminal networks

A four terminal network is symmetrical if the network remains unchanged when port designations 1 and 2 are interchanged. This implies equal open circuit transfer impedance's and equal open circuit input impedance's.

$z_{11}=z_{22}$ $z_{12}=z_{21}$ $y_{11}=y_{22}$ $y_{12}=y_{21}$
$\Delta_h=1$ $h_{12}=-h_{21}$ $\Delta_g=1$ $g_{12}=-g_{21}$
$\Delta_T=1$ $A=D$ $\Delta_t=1$ $a=d$

Problem 108 Show that the terminated four terminal network input z is

$$z_{in}=\frac{V_1}{I_1}=z_{11}-\frac{z_{12}z_{21}}{z_{22}+z_L}$$

1.3 Transfer Functions

In practice a four terminal network is terminated at both ends. It is driven by a source with finite impedance z_S and terminated with a load impedance z_L (Figure 102a). Also the four terminal network may be terminated at only one end. Then the source is a high impedance constant current generator.

Figure 102 Terminated four terminal network

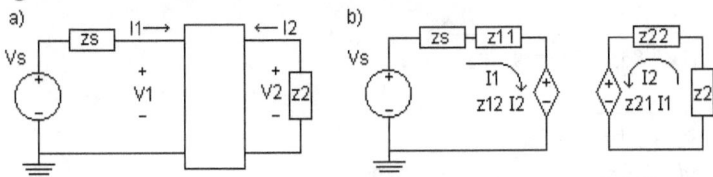

Four terminal network terminated at one end (I_1 in, V_2 out).

(14) $-z_{21}I_1 = (z_{22} + z_L)I_2 \qquad z_2 = z_L \qquad v = Iz \; here$

(15a) $\dfrac{I_2}{I_1} = -\dfrac{z_{21}}{z_{22} + z_L}$
(15b) $\dfrac{V_2}{I_1} = \dfrac{-I_2 z_L}{I_1} = \dfrac{z_{21} z_L}{z_{22} + z_L}$

Four terminal network terminated at both ends (V_1 in, V_2 out).
We start with the port mesh equations (Figure 102b).

(16a) $V_S = (z_{11} + z_S)I_1 + z_{12}I_2$

(16b) $0 = z_{21}I_1 + (z_{22} + z_L)I_2$

response currents :

(17a) $\dfrac{I_1}{V_S} = \dfrac{(z_{22} + z_L)}{(z_{11} + z_S)(z_{22} + z_L) - z_{12}z_{21}} = \dfrac{z_{22} + z_L}{\Delta_{ZT}}$

(17b) $\dfrac{I_2}{V_S} = \dfrac{-z_{21}}{(z_{11} + z_S)(z_{22} + z_L) - z_{12}z_{21}} = -\dfrac{z_{21}}{\Delta_{ZT}}$

(18) $\dfrac{V_2}{V_S} = \dfrac{-I_2 z_L}{V_S} = \dfrac{z_{21} z_L}{(z_{11} + z_S)(z_{22} + z_L) - z_{12}z_{21}}$

Problem 109 Show that the terminated four terminal network input z is

$$z_{in} = \dfrac{V_1}{I_1} = h_{11} - \dfrac{h_{12}h_{21}z_L}{1 + h_{22}z_L}$$

Table 103 Transfer Functions

$$\frac{V_2}{V_S} = \frac{z_{21}z_L}{(z_{11}+z_S)(z_{22}+z_L)-z_{12}z_{21}}$$

$$\frac{V_2}{V_S} = \frac{y_{21}z_L}{y_{12}y_{21}z_S z_L - (1+y_{11}z_S)(1+y_{22}z_L)}$$

$$\frac{V_2}{V_S} = \frac{-h_{21}z_L}{(h_{11}+z_S)(1+h_{22}z_L)-h_{12}h_{21}z_L}$$

$$\frac{V_2}{V_S} = \frac{g_{21}z_L}{(1+g_{11}z_S)(g_{22}+z_L)-g_{12}g_{21}z_S}$$

$$\frac{V_2}{V_S} = \frac{z_L}{(A+Cz_S)z_L + B + Dz_S}$$

$$\frac{V_2}{V_S} = \frac{(ad-bc)z_L}{b+az_S+dz_L+cz_S z_L}$$

Example 105 Calculating a Transfer Function

The calculation of the transfer function T(p) for this low pass filter starts by calculating the z parameters for the LC four terminal network, and then substituting them into the formula for T(p) in Table 103.

$$z_S = z_L = R \qquad z_{11} = z_{22} = pL + \frac{1}{pC} \qquad z_{12} = z_{21} = \frac{1}{pC}$$

$$T(p) = \frac{V_2}{V_S} = \frac{z_{21}z_L}{(z_{11}+z_S)(z_{22}+z_L)-z_{12}z_{21}}$$

$$T(p) = \frac{V_2}{V_S} = \frac{\dfrac{1}{pC}R}{\left[R+pL+\dfrac{1}{pC}\right]^2 - \left[\dfrac{1}{pC}\right]^2} = \frac{R}{pC}\frac{1}{(R+pL)^2 + \dfrac{2}{pC}(R+pL)}$$

$$= \frac{R}{pC}\frac{1}{(R+pL)\left[pL+R+\dfrac{2}{pC}\right]} = \frac{R}{(R+pL)(LCp^2 + RCp + 2)}$$

1.4 Thevenin Equivalent Circuit

The Thevenin output impedance z_T and voltage v_T provide an equivalent circuit driving the load impedance z_L. If V_2 is replaced by a test source V_y then I_2 becomes I_y. If we solve the mesh equations for I_y and recast the solution we will find the expressions for z_T and v_T.

(19a) $\quad V_S = (z_{11} + z_S)I_1 + z_{12}I_y$

(19b) $\quad V_y = z_{21}I_1 + z_{22}I_y$

$$V_y = z_{21}\frac{V_S - z_{12}I_y}{z_{11} + z_S} + z_{22}I_y$$

$$V_y = \left(z_{22} - \frac{z_{12}z_{21}}{(z_{11} + z_S)}\right)I_y + \frac{z_{21}}{(z_{11} + z_S)}V_S$$

$$V_y = \qquad\qquad z_T I_y + v_T$$

(20a) $\quad z_T = z_{22} - \dfrac{z_{12}z_{21}}{z_{11} + z_S}$ $\qquad$ (20b) $\quad v_T = \dfrac{z_{21}}{(z_{11} + z_S)}V_S$

Table 104 Thevenin Parameters

parameter	z_T	v_T
z	$z_{22} - \dfrac{z_{12}z_{21}}{z_{11} + z_S}$	$V_S \dfrac{z_{21}}{z_{11} + z_S}$
y	$\dfrac{1 + y_{11}z_S}{y_{22} + \Delta_y z_S}$	$V_S \dfrac{-y_{21}z_S}{y_{22} + \Delta_y z_S}$
$ABCD$	$\dfrac{B + Dz_S}{A + Cz_S}$	$V_S \dfrac{1}{A + Cz_S}$
$abcd$	$\dfrac{b + az_S}{d + cz_S}$	$V_S \dfrac{\Delta_t}{d + cz_S}$
h	$\dfrac{h_{11} + z_S}{h_{22}z_S + \Delta_h}$	$V_S \dfrac{-h_{21}}{h_{22}z_S + \Delta_h}$
g	$g_{22} - \dfrac{g_{12}g_{21}z_S}{1 + g_{11}z_S}$	$V_S \dfrac{g_{21}}{1 + g_{11}z_S}$

1.5 Interconnected Four terminal networks

There are five ways to interconnect the inputs and outputs of four terminal networks: 1. series-series   2. series-parallel   3. parallel-series   4. parallel-parallel   5. cascade

Figure 103 Interconnected four terminal networks

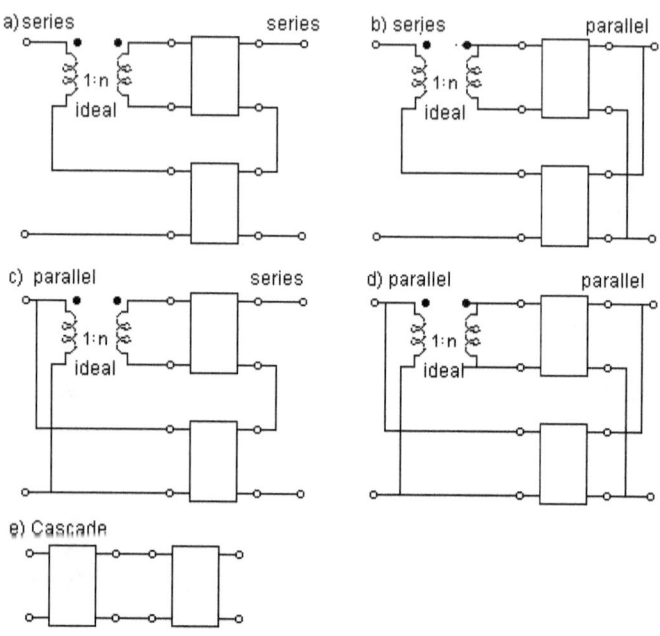

1. series-series (add z_{ij})

$$\begin{vmatrix} V_{1a} \\ V_{2a} \end{vmatrix} = \begin{vmatrix} Z_{11} & Z_{12} \\ Z_{21} & Z_{22} \end{vmatrix} \begin{vmatrix} I_{1a} \\ I_{2a} \end{vmatrix} \qquad \begin{aligned} V_{1a} &= Z_{11}I_{1a} + Z_{12}I_{2a} \\ V_{2a} &= Z_{21}I_{1a} + Z_{22}I_{2a} \end{aligned}$$

$$\begin{vmatrix} V_{1b} \\ V_{2b} \end{vmatrix} = \begin{vmatrix} z_{11} & z_{12} \\ z_{21} & z_{22} \end{vmatrix} \begin{vmatrix} I_{1b} \\ I_{2b} \end{vmatrix} \qquad \begin{aligned} V_{1b} &= z_{11}I_{1b} + z_{12}I_{2b} \\ V_{2b} &= z_{21}I_{1b} + z_{22}I_{2b} \end{aligned}$$

$$V_1 = V_{1a} + V_{1b} \qquad\qquad V_2 = V_{2a} + V_{2b}$$

$$I_1 = I_{1a} = I_{1b} \qquad\qquad I_2 = I_{2a} = I_{2b}$$

$$V_1 = V_{1a} + V_{1b} = Z_{11}I_{1a} + Z_{12}I_{2a} + z_{11}I_{1b} + z_{12}I_{2b}$$

$$V_2 = V_{2a} + V_{2b} = Z_{21}I_{1a} + Z_{22}I_{2a} + z_{21}I_{1b} + z_{22}I_{2b}$$

$$\boxed{\begin{aligned} V_1 &= (Z_{11} + z_{11})I_1 + (Z_{12} + z_{12})I_2 \\ V_2 &= (Z_{21} + z_{21})I_1 + (Z_{22} + z_{22})I_2 \end{aligned}}$$

2. series-parallel (add h_{ij})

$$\begin{vmatrix} V_{1a} \\ I_{2a} \end{vmatrix} = \begin{vmatrix} H_{11} & H_{12} \\ H_{21} & H_{22} \end{vmatrix} \begin{vmatrix} I_{1a} \\ V_{2a} \end{vmatrix} \quad \begin{aligned} V_{1a} &= H_{11}I_{1a} + H_{12}V_{2a} \\ I_{2a} &= H_{21}I_{1a} + H_{22}V_{2a} \end{aligned}$$

$$\begin{vmatrix} V_{1b} \\ I_{2b} \end{vmatrix} = \begin{vmatrix} h_{11} & h_{12} \\ h_{21} & h_{22} \end{vmatrix} \begin{vmatrix} I_{1b} \\ V_{2b} \end{vmatrix} \quad \begin{aligned} V_{1b} &= h_{11}I_{1b} + h_{12}V_{2b} \\ I_{2b} &= h_{21}I_{1b} + h_{22}V_{2b} \end{aligned}$$

$$\begin{aligned} V_1 &= V_{1a} + V_{1b} & V_2 &= V_{2a} = V_{2b} \\ I_1 &= I_{1a} = I_{1b} & I_2 &= I_{2a} + I_{2b} \end{aligned}$$

$$V_1 = V_{1a} + V_{1b} = H_{11}I_{1a} + H_{12}V_{2a} + h_{11}I_{1b} + h_{12}V_{2b}$$
$$I_2 = I_{2a} + I_{2b} = H_{21}I_{1a} + H_{22}V_{2a} + h_{21}I_{1b} + h_{22}V_{2b}$$

$$\boxed{\begin{aligned} V_1 &= (H_{11} + h_{11})I_1 + (H_{12} + h_{12})V_2 \\ I_2 &= (H_{21} + h_{21})I_1 + (H_{22} + h_{22})V_2 \end{aligned}}$$

3. parallel-series (add gij)

$$\begin{vmatrix} I_{1a} \\ V_{2a} \end{vmatrix} = \begin{vmatrix} G_{11} & G_{12} \\ G_{21} & G_{22} \end{vmatrix} \begin{vmatrix} V_{1a} \\ I_{2a} \end{vmatrix} \quad \begin{aligned} I_{1a} &= G_{11}V_{1a} + G_{12}I_{2a} \\ V_{2a} &= G_{21}V_{1a} + G_{22}I_{2a} \end{aligned}$$

$$\begin{vmatrix} I_{1b} \\ V_{2b} \end{vmatrix} = \begin{vmatrix} g_{11} & g_{12} \\ g_{21} & g_{22} \end{vmatrix} \begin{vmatrix} V_{1b} \\ I_{2b} \end{vmatrix} \quad \begin{aligned} I_{1b} &= g_{11}V_{1b} + g_{12}I_{2b} \\ V_{2b} &= g_{21}V_{1b} + g_{22}I_{2b} \end{aligned}$$

$$\begin{aligned} V_1 &= V_{1a} = V_{1b} & V_2 &= V_{2a} + V_{2b} \\ I_1 &= I_{1a} + I_{1b} & I_2 &= I_{2a} = I_{2b} \end{aligned}$$

$$I_1 = I_{1a} + I_{1b} = G_{11}V_{1a} + G_{12}I_{2a} + g_{11}V_{1b} + g_{12}I_{2b}$$
$$V_2 = V_{2a} + V_{2b} = G_{21}V_{1a} + G_{22}I_{2a} + g_{21}V_{1b} + g_{22}I_{2b}$$

$$\boxed{\begin{aligned} I_1 &= (G_{11} + g_{11})V_1 + (G_{12} + g_{12})I_2 \\ V_2 &= (G_{21} + g_{21})V_1 + (G_{22} + g_{22})I_2 \end{aligned}}$$

4. parallel-parallel (add y_{ij})

$$\begin{vmatrix} I_{1a} \\ I_{2a} \end{vmatrix} = \begin{vmatrix} Y_{11} & Y_{12} \\ Y_{21} & Y_{22} \end{vmatrix} \begin{vmatrix} V_{1a} \\ V_{2a} \end{vmatrix} \quad \begin{aligned} I_{1a} &= Y_{11}V_{1a} + Y_{12}V_{2a} \\ I_{2a} &= Y_{21}V_{1a} + Y_{22}V_{2a} \end{aligned}$$

$$\begin{vmatrix} I_{1b} \\ I_{2b} \end{vmatrix} = \begin{vmatrix} y_{11} & y_{12} \\ y_{21} & y_{22} \end{vmatrix} \begin{vmatrix} V_{1b} \\ V_{2b} \end{vmatrix} \quad \begin{aligned} I_{1b} &= y_{11}V_{1b} + y_{12}V_{2b} \\ I_{2b} &= y_{21}V_{1b} + y_{22}V_{2b} \end{aligned}$$

$$V_1 = V_{1a} = V_{1b} \qquad V_2 = V_{2a} = V_{2b}$$
$$I_1 = I_{1a} + I_{1b} \qquad I_2 = I_{2a} + I_{2b}$$

$$I_1 = I_{1a} + I_{1b} = Y_{11}V_{1a} + Y_{12}V_{2a} + y_{11}V_{1b} + y_{12}V_{2b}$$
$$I_2 = I_{2a} + I_{2b} = Y_{21}V_{1a} + Y_{22}V_{2a} + y_{21}V_{1b} + y_{22}V_{2b}$$

$$\boxed{\begin{aligned} I_1 &= (Y_{11} + y_{11})V_1 + (Y_{12} + y_{12})V_2 \\ I_2 &= (Y_{21} + y_{21})V_1 + (Y_{22} + y_{22})V_2 \end{aligned}}$$

5. cascade (multiply ABCD) This interconnection is different, because the two matrices are multiplied instead of being added.

$$\begin{vmatrix} V_{1a} \\ I_{1a} \end{vmatrix} = \begin{vmatrix} A_1 & B_1 \\ C_1 & D_1 \end{vmatrix} \begin{vmatrix} V_{2a} \\ -I_{2a} \end{vmatrix} \quad \begin{aligned} V_{1a} &= A_1 V_{2a} - B_1 I_{2a} \\ I_{1a} &= C_1 V_{2a} - D_1 I_{2a} \end{aligned}$$

$$\begin{vmatrix} V_{1b} \\ I_{1b} \end{vmatrix} = \begin{vmatrix} A_2 & B_2 \\ C_2 & D_2 \end{vmatrix} \begin{vmatrix} V_{2b} \\ -I_{2b} \end{vmatrix} \quad \begin{aligned} V_{1b} &= A_2 V_{2b} - B_2 I_{2b} \\ I_{1b} &= C_2 V_{2b} - D_2 I_{2b} \end{aligned}$$

$$V_1 = V_{1a} \; V_2 = V_{2b} \qquad V_{2a} = V_{1b}$$
$$I_1 = I_{1a} \; I_2 = I_{2b} \qquad I_{2a} = -I_{1b}$$

$$Since\, V_{2a} = V_{1b}\, and\, I_{2a} = -I_{1b}$$
$$V_1 = V_{1a} = A_1 V_{2a} - B_1 I_{2a} = A_1 V_{1b} - B_1 I_{2a}$$
$$V_1 = A_1(A_2 V_{2b} - B_2 I_{2b}) - B_1(-C_2 V_{2b} + D_2 I_{2b})$$
$$\boxed{V_1 = (A_1 A_2 + B_1 C_2)V_2 - (A_1 B_2 + B_1 D_2)I_2}$$

and

$$I_1 = I_{1a} = C_1 V_{2a} - D_1 I_{2a} = C_1 V_{1b} + D_1 I_{2a}$$
$$I_1 = C_1(A_2 V_{2b} - B_2 I_{2b}) - D_1(-C_2 V_{2b} + D_2 I_{2b})$$
$$\boxed{I_1 = (C_1 A_2 + D_1 C_2)V_2 - (C_1 B_2 + D_1 D_2)I_2}$$

Example 106 Parallel-parallel Connection

Compare this method to starting from scratch with a nodal analysis.

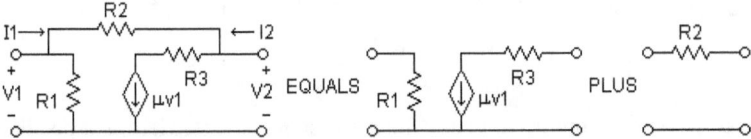

Parallel-inputs and parallel outputs imply using y parameters.

1. y parameters for the amplifier four terminal network:

$$y_{11} = \frac{I_1}{V_1}\bigg|_{V_2=0} = g_1 \qquad y_{12} = \frac{I_1}{V_2}\bigg|_{V_1=0} = 0$$

$$y_{21} = \frac{I_2}{V_1}\bigg|_{V_2=0} = \mu g_3 \qquad y_{22} = \frac{I_2}{V_2}\bigg|_{V_1=0} = g_3$$

2. y parameters for the feedback resistor four terminal network:

$$y_{11} = \frac{I_1}{V_1}\bigg|_{V_2=0} = g_2 \qquad y_{12} = \frac{I_1}{V_2}\bigg|_{V_1=0} = -g_2$$

$$y_{21} = \frac{I_2}{V_1}\bigg|_{V_2=0} = -g_2 \qquad y_{22} = \frac{I_2}{V_2}\bigg|_{V_1=0} = g_2$$

3. y parameters for the feedback amplifier four terminal network:

$$y_{11} = \frac{I_1}{V_1}\bigg|_{V_2=0} = g_1 + g_2 \qquad y_{12} = \frac{I_1}{V_2}\bigg|_{V_1=0} = -g_2$$

$$y_{21} = \frac{I_2}{V_1}\bigg|_{V_2=0} = -g_2 + \mu g_3 \qquad y_{22} = \frac{I_2}{V_2}\bigg|_{V_1=0} = g_2 + g_3$$

Example 107 Cascade Connection

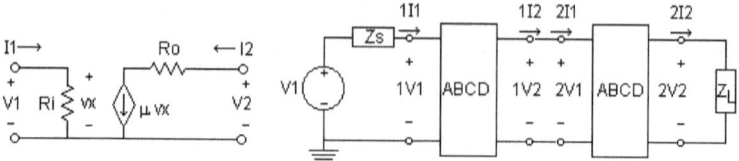

We build a two-stage amplifier by cascading two single stages. Here is a detailed analysis.

$$_2V_2 = \frac{z_L}{z_L + R_{02}}\mu_2(v_{x2}) = \frac{z_L}{z_L + R_{02}}\mu_2(_2I_2R_{i2}) = \frac{z_L}{z_L + R_{02}}\mu_2 R_{i2}(_1I_2)$$

$$= \frac{z_L}{z_L + R_{02}}\mu_2 R_{i2}\left(\frac{\mu_1 v_{x1}}{R_{i2} + R_{01}}\right) = \frac{z_L}{z_L + R_{02}}\mu_2\frac{R_{i2}}{R_{i2} + R_{01}}\mu_1(v_{x1})$$

$$= \frac{z_L}{z_L + R_{02}}\mu_2\frac{R_{i2}}{R_{i2} + R_{01}}\mu_1(_1I_1R_{i1}) = \frac{z_L}{z_L + R_{02}}\mu_2\frac{R_{i2}}{R_{i2} + R_{01}}\mu_1 R_{i1}(_1I_1)$$

$$= \frac{z_L}{z_L + R_{02}}\mu_2\frac{R_{i2}}{R_{i2} + R_{01}}\mu_1 R_{i1}\left(\frac{V_s}{R_i + z_s}\right)$$

$$= \frac{z_L}{z_L + R_{02}}\mu_2\frac{R_{i2}}{R_{i2} + R_{01}}\mu_1\frac{R_{i1}}{R_{i1} + z_s}V_s$$

Solve by using four terminal networks. First we need the amplifier's ABCD parameters.

$$V_1 = AV_2 - BI_2$$
$$I_1 = CV_2 - DI_2$$

AIf $I_2 = 0$ then $V_2 = \mu V_1$.
BIf $V_2 = 0$ then $I_2 = \mu V_1/R_o$
CIf $I_2 = 0$ then $V_2 = \mu V_1 = \mu R_i I_1$
D If $V_2 = 0$ then $I_2 = \mu V_1/R_o = \mu R_i I_1/R_o$

$$A = \left.\frac{V_1}{V_2}\right|_{I_2=0} = -\frac{1}{\mu} \qquad B = \left.-\frac{V_1}{I_2}\right|_{V_2=0} = -\frac{R_0}{\mu}$$

$$C = \left.\frac{I_1}{V_2}\right|_{I_2=0} = -\frac{1}{\mu R_i} \qquad D = \left.-\frac{I_1}{I_2}\right|_{V_2=0} = -\frac{R_0}{\mu R_i}$$

I_2 and V_2 of the first stage equal the second stage I_1 and V_1. So we could substitute the second stage ABCD equations into those of the first stage to find the two-stage ABCD parameters. A better way is to multiply two the two ABCD T matrices.

$$T = \begin{vmatrix} A & B \\ C & D \end{vmatrix} = -\frac{1}{\mu} \begin{vmatrix} 1 & R_0 \\ \dfrac{1}{R_i} & R_i \end{vmatrix}$$

The product of two identical T matrices is

$$T = \begin{vmatrix} A & B \\ C & D \end{vmatrix}\begin{vmatrix} A & B \\ C & D \end{vmatrix} = \begin{vmatrix} AA+BC & AB+BD \\ CA+DC & CB+DD \end{vmatrix}$$

Substitute values for A, B, C, and D to get the two-stage ABCD matrix T^2.

$$T^2 = \frac{1}{\mu^2}\begin{vmatrix} 1 & R_0 \\ \dfrac{1}{R_i} & R_i \end{vmatrix}\begin{vmatrix} 1 & R_0 \\ \dfrac{1}{R_i} & R_i \end{vmatrix} = \frac{1}{\mu^2}\begin{vmatrix} 1+\dfrac{R_0}{R_i} & R_0+\dfrac{R_0^2}{R_i} \\ \dfrac{1}{R_i}+\dfrac{R_0}{R_i^2} & R_0+\dfrac{R_0^2}{R_i^2} \end{vmatrix} = \frac{1}{\mu^2}\left(1+\dfrac{R_0}{R_i}\right)\begin{vmatrix} 1 & R_0 \\ \dfrac{1}{R_i} & R_i \end{vmatrix}$$

The transfer function using ABCD parameters (Table 103 Transfer Functions

$$\frac{V_2}{V_S} = \frac{z_L}{(A+Cz_S)z_L + B + Dz_S}$$

if $k = \dfrac{1}{\mu^2}\left(1+\dfrac{R_0}{R_i}\right)$ then the A, B, C, and D of the T^2 matrix are

$$A = k, \quad B = kR_0, \quad C = k\frac{1}{R_i}, \quad D = k\frac{R_0}{R_i}$$

$$\frac{V_2}{V_S} = \frac{z_L}{(k+k\dfrac{1}{R_i}z_S)z_L + kR_0 + k\dfrac{R_0}{R_i}z_S} = \frac{1}{k}\frac{z_L}{\left(1+\dfrac{z_S}{R_i}\right)z_L + R_0\left(1+\dfrac{z_S}{R_i}\right)}$$

$$= \frac{1}{\dfrac{1}{\mu^2}\left(1+\dfrac{R_0}{R_i}\right)\left(1+\dfrac{z_S}{R_i}\right)}\frac{z_L}{z_L+R_0} = \mu^2\frac{R_i}{R_i+z_S}\frac{R_i}{R_i+R_0}\frac{z_L}{z_L+R_0}$$

2 Ladder Filter Design

The Bell Telephone Laboratories ladder filter sections[4] constant k, and m derived equations are developed in a straightforward manner. The general *propagation function* and *image impedance* for sections are derived. The ratio of output voltage to input voltage is $e^{-\gamma}$. Then T and π *image impedance* equations are developed. Cascading allows a designer to assemble filters with a wide variety of transfer functions.

2.1 Propagation Function

A propagation function is the ratio of a network output to the network input. The immediate goal is finding the propagation function of an infinite cascade of identical two port networks (Figure 201).

Figure 201 Cascade of Identical Two Port Networks

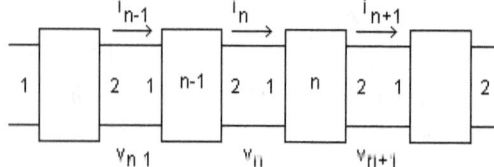

Equations 1 and 2 are the two port admittance equations of two ports n−1 and n (Figure 201, Table 101 pg 3).

$(1a)\quad i_{n-1} = y_{11}v_{n-1} + y_{12}v_n$ $\qquad (2a)\quad i_n = y_{11}v_n + y_{12}v_{n+1}$

$(1b)\quad -i_n = y_{21}v_{n-1} + y_{22}v_n$ $\qquad (2b)\quad -i_{n+1} = y_{21}v_n + y_{22}v_{n+1}$

The current $-i_n$ out of two port n−1 equals current i_n into two port n. Therefore current is eliminated when equation 1b is set equal to equation 2a.

$(3a)\quad -I_2 \text{ of } n-1 = I_1 \text{ of } n \quad \rightarrow \quad -i_n \text{ of } n-1 = i_n \text{ of } n$

$$- y_{21}v_{n-1} - y_{22}v_n = y_{11}v_n + y_{12}v_{n+1}$$

$$0 = y_{21}v_{n-1} + y_{22}v_n + y_{11}v_n + y_{12}v_{n+1}$$

$$0 = +y_{21}v_{n-1} + (y_{11} + y_{22})v_n + y_{12}v_{n+1}$$

---

[4] Zobel, O. J.,*Theory and Design of Uniform and Composite Electric Wave Filters*, Bell Systems Technical Journal, Vol. 2 (1923), pp. 1–46.

Cast equation 3a into the form of a difference equation by dividing by y_{12}.

$$(3b) \quad 0 = v_{n-1} + \left(\frac{y_{11} + y_{22}}{y_{12}} \right) v_n + v_{n+1} \qquad \rightarrow \qquad y_{12} = y_{21}$$

Equations 4 and 5 are the impedance equations of two ports n–1 and n (Table 101 pg 3).

$$(4a) \quad v_{n-1} = z_{11} i_{n-1} - z_{12} i_n \qquad\qquad (5a) \quad v_n = z_{11} i_n - z_{12} i_{n+1}$$
$$(4b) \quad v_n = z_{21} i_{n-1} - z_{22} i_n \qquad\qquad (5b) \quad v_{n+1} = z_{21} i_n - z_{22} i_{n+1}$$

The associated current difference equation is created when voltage is eliminated. The voltage v_n out of two port n–1 equals voltage v_n into two port n. Therefore voltage is eliminated when equation 4b is set equal to equation 5a.

$$(6a) \quad V_2 \text{ of } n-1 = V_1 \text{ of } n \quad \rightarrow \quad v_n \text{ of } n-1 = v_n \text{ of } n$$
$$z_{21} i_{n-1} - z_{22} i_n = z_{11} i_n - z_{12} i_{n+1}$$
$$0 = -z_{21} i_{n-1} + z_{22} i_n + z_{11} i_n - z_{12} i_{n+1}$$
$$0 = +z_{21} i_{n-1} - (z_{11} + z_{22}) i_n + z_{12} i_{n+1}$$

Cast equation 6a into the form of a difference equation by dividing by z_{12}.

$$(6b) \quad 0 = i_{n-1} - \left(\frac{z_{11} + z_{22}}{z_{12}} \right) i_n + i_{n+1} \qquad \rightarrow \qquad z_{12} = z_{21}$$

Find a common expression for the y's and z's (Table 102 page 10).

$$(7a) \quad \frac{y_{11} + y_{22}}{y_{12}} = \frac{\dfrac{z_{22}}{\Delta_z} + \dfrac{z_{11}}{\Delta_z}}{-\dfrac{z_{12}}{\Delta_z}} = \frac{\dfrac{D}{B} + \dfrac{A}{B}}{-\dfrac{1}{B}} = -(A + D)$$

$$(7b) \quad \frac{y_{11} + y_{22}}{y_{12}} = -\frac{z_{11} + z_{22}}{z_{12}} = -(A + D)$$

Substitute 7b into 3b and 6b to eliminate the z_{ij} and y_{ij}.
$$(8a) \quad 0 = v_{n-1} - (A + D) v_n + v_{n+1}$$
$$(8b) \quad 0 = i_{n-1} - (A + D) i_n + i_{n+1}$$

Analog Filter Design

Assume solution 9 for equations 8 that has the exponential form, because exponential forms are solutions to *difference equations* such as equations 8. Let γ be the *propagation constant.*

(9a) $v_n = a_1 e^{-\gamma m}$

(9b) $i_n = b_1 e^{-\gamma m}$

The method used to solve equations 8 is an example of a *heuristic method.* Heuristic is a somewhat elegant name for judicious guessing of the form of the solution.

Form v_{n-1}, v_{n+1}, and voltage ratios.

(10a) $v_{n-1} = a_1 e^{-\gamma(n-1)} = e^{+\gamma} v_n$

(10b) $v_n = a_1 e^{-\gamma(n)}$

(10c) $v_{n+1} = a_1 e^{-\gamma(n+1)} = e^{-\gamma} v_n$

(10d) $\dfrac{v_n}{v_{n-1}} = \dfrac{v_{n+1}}{v_n} = e^{-\gamma}$ *and by the same process* $\dfrac{i_n}{i_{n-1}} = \dfrac{i_{n+1}}{i_n} = e^{-\gamma}$

Substitute (10b) into (8a)

(11a) $0 = v_{n-1} - (A+D)v_n + v_{n+1} = e^{+\gamma}v_n - (A+D)v_n + e^{-\gamma}v_n$

$0 = [e^{+\gamma} - (A+D) + e^{-\gamma}]v_n$

(11b) $0 = i_{n-1} - (A+D)i_n + i_{n+1} = e^{+\gamma}i_n - (A+D)i_n + e^{-\gamma}i_n$

$0 = [e^{+\gamma} - (A+D) + e^{-\gamma}]i_n$

Solve equations 11 for e^{γ}.

(12a) $i_n \neq 0$ $v_n \neq 0$ *so* $[e^{+\gamma} - (A+D) + e^{-\gamma}] = 0$

$e^{+\gamma} + e^{-\gamma} = A+D$

$\cosh \gamma = \dfrac{A+D}{2}$

(12b) $\sinh^2 \gamma = \cosh^2 \gamma - 1 = \left(\dfrac{A+D}{2}\right)^2 - 1$

(12c) $\sinh \gamma = \dfrac{\sqrt{(A+D)^2 - 4}}{2}$

Use the definitions of cosh γ and sinh γ to find an expression for e^γ.

(13) $\quad \cosh \gamma \pm \sinh \gamma = \dfrac{e^{+\gamma} + e^{-\gamma}}{2} \pm \dfrac{e^{+\gamma} - e^{-\gamma}}{2} = e^{\pm \gamma}$

$$e^{\pm \gamma} = \frac{(A+D) \pm \sqrt{(A+D)^2 - 4}}{2}$$

Since A=D for a symmetric network e^γ is simplified.

(14a) $\quad \dfrac{v_{out}}{v_{in}} = \dfrac{i_{out}}{i_{in}} = e^{-\gamma}{}_{symettric} = \dfrac{(A+D) - \sqrt{(A+D)^2 - 4}}{2} = \dfrac{(2A) - \sqrt{(2A)^2 - 4}}{2}$

(14b) $\quad T_{out}(p) = \dfrac{v_{out}}{v_{in}} = A - \sqrt{A^2 - 1}$

(14c) $\quad T_{in}(p) = \dfrac{v_{in}}{v_{out}} = \dfrac{1}{A - \sqrt{A^2 - 1}} \times \dfrac{A + \sqrt{A^2 - 1}}{A + \sqrt{A^2 - 1}} = A + \sqrt{A^2 - 1}$

The transfer functions are the *image impedance transfer functions*, because an infinite chain is being analyzed (Figure 201).

Let z_{in} be the input impedance of the chain. If the infinite chain is cut into two parts at node n (Figure 201) the part to the right is still infinite and its input z is still z_{in}. Throw away the right hand part and replace it with z_{in} connected across the end of the left hand part. The left hand part will not know that the right hand part was replaced by the impedance z_{in}. This creates a finite chain terminated by z_{in}, which behaves as if it is an infinite chain.

A finite cascade is terminated in the image impedance calculated in the next section (Section 2.2). The *image impedance terminated* finite cascade behaves as an infinite chain.

Figure 201 Cascade of Identical Networks

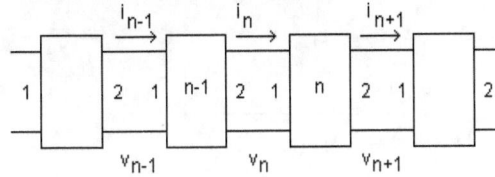

23

2.2 T and π Image Impedance

The input impedance z_1 of a network depends upon the terminating impedance z_2 (Figure 202). The question is – what value of the terminating impedance will make the input impedance z_1 equal to z_2? Apply the two port network represented by the ABCD parameters, because the input variables depend on the output variables. ABCD emphasizes input to output transmission properties (Table 102 page 10).

$$(15a) \quad v_1 = Av_2 + Bi_2$$
$$(15b) \quad i_1 = Cv_2 + Di_2$$

Figure 202

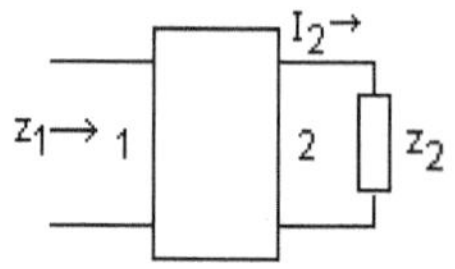

The input impedance $z_1 = v_1/i_1$. Let z_1 represent the *image* impedance of a two port.

$$(16a) \quad v_2 = i_2 z_I \qquad z_1 = z_2 = z_I$$

$$(16b) \quad z_1 = \frac{v_1}{i_1} = \frac{Av_2 + Bi_2}{Cv_2 + Di_2}$$

$$(16c) \quad z_I = \frac{Ai_2 z_I + Bi_2}{Ci_2 z_I + Di_2} = \frac{Az_I + B}{Cz_I + D}$$

$$(16d) \quad z_I(Cz_I + D) = Az_I + B \quad \rightarrow \quad Cz_I{}^2 + (D-A)z_I - B = 0$$

The two ports to be used in the filters have to be symmetric networks where A=D, because the Bell Telephone Laboratories ladder filters are assembled from symmetric sections.

$$(16e) \quad z_I = \frac{-(D-A) \pm \sqrt{(D-A)^2 + 4BC}}{2C} = \sqrt{\frac{B}{C}}$$

The image impedances expressed in terms of z's and y's (Table 102 page 10).

$$(17a) \quad z_I{}^2 = \frac{B}{C} = B\frac{1}{C} = B\frac{1}{C}\frac{\Delta_z}{z_{12}} \quad z_{12} = \Delta_z = z_{11}z_{22} - z_{12}{}^2$$

$$(17b) \quad z_I{}^2 = \frac{B}{C} = B\frac{1}{C} = B\frac{1}{C} = \frac{1}{-y_{12}} \times \frac{-y_{12}}{\Delta_y} = \frac{1}{\Delta_y} = \frac{1}{y_{11}y_{22} - y_{12}{}^2}$$

While we are at it consider z_I in terms of an open circuit load ($i_2=0$) and a short circuit load ($v_2=0$).

$$(18a) \quad z_{1open} = \left.\frac{v_1}{i_1}\right|_{i_2=0} = \left.\frac{Av_2 + Bi_2}{Cv_2 + Di_2}\right|_{i_2=0} = \frac{Av_2}{Cv_2} = \frac{A}{C}$$

$$(18b) \quad z_{1short} = \left.\frac{v_1}{i_1}\right|_{v_2=0} = \left.\frac{Av_2 + Bi_2}{Cv_2 + Di_2}\right|_{v_2=0} = \frac{Bi_2}{Di_2} = \frac{B}{D}$$

$$(18c) \quad z_{1open}z_{1short} = \frac{A}{C}\frac{B}{D} = \frac{B}{C} = z_I{}^2$$

Bell Telephone Laboratories ladder filters are assemblies of T and, or π sections. Calculate the image impedance of the T and π sections.

Figure 203a T Filter Section - any type

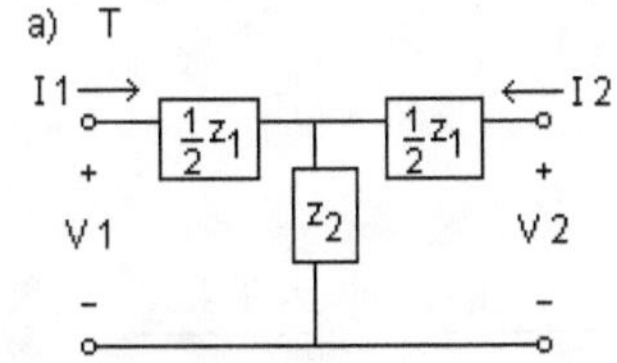

a) T

T section -

$$(19a) \quad z_{11} = z_{22} = \left.\frac{v_1}{i_1}\right|_{i_2=0} = \frac{z_1}{2} + z_2 \qquad z_{12} = \left.\frac{v_2}{i_1}\right|_{i_2=0} = z_2$$

$$(17a) \quad z_{IT}{}^2 = z_{11}z_{22} - z_{12}{}^2$$

$$(19b) \quad z_{IT}{}^2 = z_{11}z_{22} - z_{12}{}^2 = \left(\frac{z_1}{2} + z_2\right)^2 - z_2{}^2 = \frac{z_1{}^2}{4} + z_1z_2 + z_2{}^2 - z_2{}^2$$

$$= \frac{z_1{}^2}{4} + z_1z_2$$

$$(19c) \quad z_{IT} = \sqrt{\tfrac{1}{4}z_1{}^2 + z_1z_2}$$

Analog Filter Design

Figure 203b π Filter Section - any type

b) pi

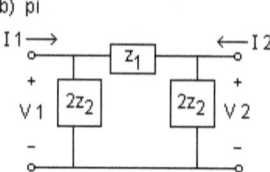

π section

(21a) $y_{11} = y_{22} = \dfrac{i_1}{v_1}\Big|_{v_2=0} = \dfrac{1}{2z_2} + \dfrac{1}{z_1}$ $\qquad y_{12} = \dfrac{i_2}{v_1}\Big|_{v_2=0} = -\dfrac{1}{z_1}$

(17b) $z_I^{\ 2} = \dfrac{1}{y_{11}y_{22} - y_{12}^{\ 2}}$

(21b) $\dfrac{1}{z_{I\pi}^{\ 2}} = y_{11}y_{22} - y_{12}^{\ 2} = \left(\dfrac{1}{2z_2} + \dfrac{1}{z_1}\right)^2 - \left(\dfrac{1}{z_1}\right)^2$

$\qquad = \dfrac{1}{4z_2^{\ 2}} + \dfrac{1}{z_1 z_2} + \left(\dfrac{1}{z_1}\right)^2 - \left(\dfrac{1}{z_1}\right)^2 = \dfrac{1}{4z_2^{\ 2}} + \dfrac{1}{z_1 z_2} = \dfrac{1}{z_2}\dfrac{z_1 + 4z_2}{4z_1 z_2}$

(21c) $z_{I\pi}^{\ 2} = z_2\dfrac{4z_1 z_2}{z_1 + 4z_2} = \dfrac{\frac{1}{4}z_1}{\frac{1}{4}z_1} \curlyvee z_2 \dfrac{4z_1 z_2}{z_1 + 4z_2} \ = \dfrac{z_1^{\ 2} z_2^{\ 2}}{\frac{1}{4}z_1^{\ 2} + z_1 z_2}$

(21d) $z_{I\pi} = \dfrac{z_1 z_2}{\sqrt{\frac{1}{4}z_1^{\ 2} + z_1 z_2}}$

Problem 201 Derive $y_{I\pi}^{\ 2} = \frac{1}{4}y_2^{\ 2} + y_1 y_2$

Problem 202 Plot $z_{ITk} = \sqrt{\dfrac{L_k}{C_k}\left(1 - \dfrac{\omega^2}{\omega_c^{\ 2}}\right)}$ *when* $\dfrac{L_k}{C_k} = 1$ from $\omega=0$ to ω_c.

Problem 203 Ref Prob 201, eqns 19c, 22b. Derive $y_{I\pi k} = \sqrt{\dfrac{C_k}{L_k}\left(1 - \dfrac{\omega^2}{\omega_c^{\ 2}}\right)}$

2.3 Constant k Filter Sections

A constant k filter section is a T or π ladder structure, which has an image impedance, a propagation function, a pass band, and an attenuation band. The low pass equations are derived here. Filter equations and schematics suitable for design are listed in the Design pages 42, 43, 44.

Observe that the T and π networks are duals.

Figure 204 Constant k Low Pass T and π sections

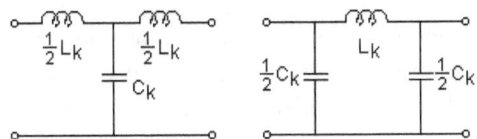

Image impedance z_I The ideal image impedance z_I is a constant R that is independent of frequency. Constant k filter sections z_I do not equal R. The low pass filter T and π image impedances are functions of frequency (equation 22c).

$(19c) \quad z_{IT} = \sqrt{\tfrac{1}{4}z_1^2 + z_1 z_2}$

$(22a) \quad \text{let } z_1 = pL_k \qquad z_2 = \dfrac{1}{pC_k} \qquad \omega_0^2 = \dfrac{1}{L_k C_k} \qquad \omega_c = 2\omega_0$

$(22b) \quad z_{ITk}^2 = \tfrac{1}{4}z_1^2 + z_1 z_2 = \tfrac{1}{4}p^2 L_k^2 + \dfrac{L_k}{C_k}$

$$= \tfrac{1}{4}p^2 L_k^2 \dfrac{\omega_0^2}{\omega_0^2} + \dfrac{L_k}{C_k}$$

$$= \tfrac{1}{4}\omega_0^2 L_k^2 \dfrac{p^2}{\omega_0^2} + \dfrac{L_k}{C_k}$$

$$= \tfrac{1}{4}\dfrac{1}{L_k C_k} L_k^2 \dfrac{p^2}{\omega_0^2} + \dfrac{L_k}{C_k}$$

$$= \tfrac{1}{4}\dfrac{L_k}{C_k}\dfrac{p^2}{\omega_0^2} + \dfrac{L_k}{C_k}$$

$$= \dfrac{L_k}{C_k}\left(1 + \dfrac{j^2\omega^2}{4\omega_0^2}\right) = \dfrac{L_k}{C_k}\left(1 - \dfrac{\omega^2}{\omega_c^2}\right)$$

$(22c)$ $z_{ITk} = \sqrt{\dfrac{L_k}{C_k}\left(1 - \dfrac{\omega^2}{\omega_c^{\,2}}\right)}$

z_{IT} changes from real to imaginary when $\omega > 2\omega_0 = \omega_{cutoff}$

$(22d)$ $\omega_{cutoff} = 2\omega_0 = \dfrac{2}{\sqrt{L_k C_k}}$

$(22e)$ $z_{ITk} = j\sqrt{\dfrac{L_k}{C_k}\left(\dfrac{\omega^2}{\omega_c^{\,2}} - 1\right)}$ $\omega > \omega_c$

Constant k components As a practical matter filters are terminated by resistors R. This is why the image impedance is replaced by resistor R as a filter source and load. The L_k and C_k are calculated using R and the cutoff frequency ω_c.

$(22f)$ $R^2 = \dfrac{L_k}{C_k}$ and $\omega_c^{\,2} = 4\omega_0^{\,2} = \dfrac{4}{L_k C_k}$ $\rightarrow$ $C_k = \dfrac{4}{\omega_c^{\,2} L_k}$

$(22g)$ $L_k = R^2 C_k = \dfrac{4R^2}{\omega_c^{\,2} L_k}$ $\rightarrow$ $L_k^{\,2} = \dfrac{4R^2}{\omega_c^{\,2}}$ $\rightarrow$ $L_k = \dfrac{2R}{\omega_c} = \dfrac{R}{\pi f_c}$

$(22h)$ $C_k = \dfrac{L_k}{R^2} = \dfrac{1}{R^2} \times \dfrac{R}{\pi f_c} = \dfrac{1}{\pi f_c R}$

Propagation Constant In general propagation constant gamma γ is a complex number $\alpha + j\beta$, where α represents attenuation and β represents phase shift.

$(12a)$ $\cosh \gamma = \dfrac{A + D}{2} = A$ $(A = D)$

$(12c)$ $\sinh \gamma = \dfrac{\sqrt{(A+D)^2 - 4}}{2} = \sqrt{A^2 - 1}$

$(23a)$ $\cosh \gamma = A = \dfrac{z_{11}}{z_{21}} = \left(\dfrac{z_1}{2} + z_2\right)\dfrac{1}{z_2} = 1 + \dfrac{z_1}{2z_2}$

$(23b)$ $2\sinh^2 \dfrac{\gamma}{2} = \cosh \gamma - 1 = \dfrac{z_1}{2z_2}$ $\rightarrow$ $\sinh\dfrac{\gamma}{2} = \sqrt{\dfrac{z_1}{4z_2}}$

Problem 204 Reference equations 14b, 23a. Show that

$(14b)$ $T_{out}(p) = 1 + \tfrac{1}{2} p^2 L_k C_k - \sqrt{p^2 L_k C_k \left(1 + \tfrac{1}{4} p^2 L_k C_k\right)}$

Pass and Attenuation Bands Expand sinh $\gamma/2$ into its real and imaginary parts.

$$(23c) \quad sinh\frac{\gamma}{2} = sinh\frac{\alpha+j\beta}{2} = sinh\frac{\alpha}{2}\cosh\frac{j\beta}{2} + \cosh\frac{\alpha}{2}\sinh\frac{j\beta}{2}$$

$$= sinh\frac{\alpha}{2}\cos\frac{\beta}{2} + j\cosh\frac{\alpha}{2}\sin\frac{\beta}{2}$$

Select values for α and β to invoke pass and attenuation bands. First observe that the constant k propagation function sinh $\gamma/2$ is imaginary.

$$(23d) \quad sinh\frac{\gamma}{2} = \sqrt{\frac{z_1}{4z_2}} = \sqrt{\frac{j^2\omega^2 L_k C_k}{4}} = \sqrt{\frac{j^2\omega^2}{4\omega_0^2}} = j\frac{\omega}{2\omega_0}$$

In the pass band $\gamma = \alpha+j\beta = 0+j\beta$. The attenuation α is zero, and the phase β ranges from 0 to π as frequency ω ranges from 0 to ω_c (24b)

$$(24a) \quad j\frac{\omega}{2\omega_0} = sinh\frac{\gamma}{2} = sinh\frac{\alpha}{2}\cos\frac{\beta}{2} + j\cosh\frac{\alpha}{2}\sin\frac{\beta}{2}$$

$$= sinh\frac{0}{2}\cos\frac{\beta}{2} + j\cosh\frac{0}{2}\sin\frac{\beta}{2} = 0\times\cos\frac{\beta}{2} + j1\times\sin\frac{\beta}{2}$$

$$= j\sin\frac{\beta}{2}$$

$$(24b) \quad \sin\frac{\beta}{2} = \frac{\omega}{2\omega_0} = \frac{\omega}{\omega_{cutoff}} \quad \rightarrow \quad \beta = 2\sin^{-1}\frac{\omega}{\omega_{cutoff}}$$

In the attenuation band $\gamma = \alpha+j\beta = \alpha+j\pi$. The attenuation α increases as frequency ω ranges from ω_c to infinity. The phase β has to be π to produce the imaginary result j cosh $\alpha/2$.

$$(25a) \quad j\frac{\omega}{2\omega_0} = sinh\frac{\gamma}{2} = sinh\frac{\alpha}{2}\cos\frac{\beta}{2} + j\cosh\frac{\alpha}{2}\sin\frac{\beta}{2}$$

$$= sinh\frac{\alpha}{2}\cos\frac{\pi}{2} + j\cosh\frac{\alpha}{2}\sin\frac{\pi}{2} = sinh\frac{\alpha}{2}\times 0 + j\cosh\frac{\alpha}{2}\times 1$$

$$= j\cosh\frac{\alpha}{2}$$

$$(25b) \quad \cosh\frac{\alpha}{2} = \frac{\omega}{2\omega_0} = \frac{\omega}{\omega_{cutoff}} \quad \rightarrow \quad \alpha = 2\cosh^{-1}\frac{\omega}{\omega_{cutoff}}$$

2.4 m Derived Filter Sections

The key to properly cascaded sections is equal image impedances, which connect the constant k and m derived sections.

$(26a)$ $\quad z_{I\,constant\,k} = z_{I\,m\,derived}$

$(26b)$ $\quad z_{ITk}^{\;2} = z_{ITm}^{\;2}$ $\qquad \rightarrow \qquad z_{1k}z_{2k} + \dfrac{z_{1k}^{\;2}}{4} = z_{1m}z_{2m} + \dfrac{z_{1m}^{\;2}}{4}$

$(26c)$ $\quad z_{1k}z_{2k} = z_{1m}z_{2m} + \dfrac{z_{1m}^{\;2} - z_{1k}^{\;2}}{4}$

$(26d)$ $\quad z_{2k} = \dfrac{z_{1m}z_{2m}}{z_{1k}} + \dfrac{z_{1m}^{\;2} - z_{1k}^{\;2}}{4z_{1k}}$

Figure 205 Constant k & m Derived Low Pass T sections

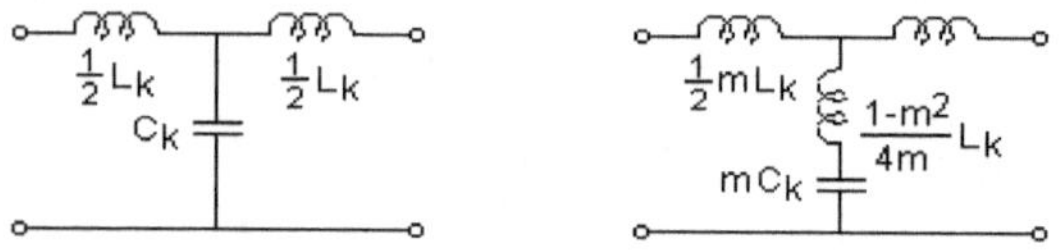

Zobel invented m derived sections when he changed $z_{1k}=L_k$ to $z_{1m}=mL_k$. Here is why.

$(27a)$ $\quad z_{1m} = mz_{1k}$

$(27b)$ $\quad z_{2k} = \dfrac{z_{1m}z_{2m}}{z_{1k}} + \dfrac{z_{1m}^{\;2} - z_{1k}^{\;2}}{4z_{1k}}$

$(27c)$ $\quad z_{2k} = \dfrac{mz_{1k}z_{2m}}{z_{1k}} + \dfrac{m^2 z_{1k}^{\;2} - z_{1k}^{\;2}}{4z_{1k}} = mz_{2m} + z_{1k}\dfrac{m^2 - 1}{4}$

$(27d)$ $\quad mz_{2m} = z_{2k} - z_{1k}\dfrac{m^2 - 1}{4}$ $\qquad \rightarrow \qquad z_{2m} = \dfrac{z_{2k}}{m} + z_{1k}\dfrac{1 - m^2}{4m}$

Changing L_k to mL_k created z_{2m} as a series LC resonant circuit (Figure 205).

(28) $\quad z_{2m} = \dfrac{z_{2k}}{m} + z_{1k}\dfrac{1 - m^2}{4m} = \dfrac{1}{mpC_k} + pL_k\dfrac{1 - m^2}{4m}$

The m derived image impedance z_{ITm} equals the constant k z_{ITk}.

$$(29) \quad z_{ITm}^2 = z_{1m}z_{2m} + \frac{z_{1m}^2}{4} = mpL_k\left(\frac{1}{mpC_k} + pL_k\frac{1-m^2}{4m}\right) + \frac{m^2p^2L_k^2}{4}$$

$$= \frac{L_k}{C_k} + \frac{p^2L_k^2}{4} = \frac{L_k}{C_k}\left(1 + \frac{p^2L_kC_k}{4}\right) = \frac{L_k}{C_k}\left(1 - \frac{\omega^2}{\omega_c^2}\right) = z_{ITk}^2$$

Equation 29 shows that the m derived cutoff frequency equals the constant k cutoff frequency (equation 22b page 27).

Figure 206 Constant k & m Derived Low Pass π sections

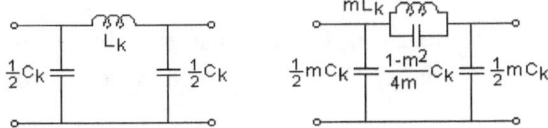

Pass and Attenuation Bands Expand sinh $\gamma/2$ into its real and imaginary parts.

$$(23c) \quad sinh\frac{\gamma}{2} = sinh\frac{\alpha + j\beta}{2} = sinh\frac{\alpha}{2}\cosh\frac{j\beta}{2} + \cosh\frac{\alpha}{2}\sinh\frac{j\beta}{2}$$

$$= sinh\frac{\alpha}{2}\cos\frac{\beta}{2} + j\cosh\frac{\alpha}{2}\sin\frac{\beta}{2}$$

Select values for α and β to invoke pass and attenuation bands. First observe that the constant k propagation function sinh $\gamma/2$ is imaginary.

$$(30a) \quad sinh\frac{\gamma}{2} = \sqrt{\frac{z_{1m}}{4z_{2m}}} = \sqrt{\frac{m^2z_{1k}}{4z_{2k} + z_{1k}(1-m^2)}}$$

$$= \sqrt{\frac{m^2 j\omega L_k}{\frac{4}{j\omega C_k} + j\omega L_k(1-m^2)}} = \sqrt{\frac{m^2 j^2\omega^2 L_kC_k}{4 + j^2\omega^2 L_kC_k(1-m^2)}}$$

$$= j\sqrt{\frac{m^2\omega^2/\omega_0^2}{4-(1-m^2)\omega^2/\omega_0^2}} = j\sqrt{\frac{m^2\omega^2 4/\omega_c^2}{4-(1-m^2)\omega^2 4/\omega_c^2}}$$

$$= j\sqrt{\frac{m^2\omega^2/\omega_c^2}{1-(1-m^2)\omega^2/\omega_c^2}}$$

where $\omega_c = 2\omega_0 \rightarrow \frac{1}{\omega_0^2} = \frac{4}{\omega_c^2}$ *and the 4's cancel*

Analog Filter Design

In the pass band $\gamma = \alpha + j\beta = 0 + j\beta$. The attenuation α is zero, and the phase β ranges from 0 to π as frequency ω ranges from 0 to ω_c (24b)

$$(31a) \quad j\sqrt{\frac{m^2\omega^2/\omega_c^2}{1-(1-m^2)\omega^2/\omega_c^2}} = \sinh\frac{\gamma}{2} = \sinh\frac{\alpha}{2}\cos\frac{\beta}{2} + j\cosh\frac{\alpha}{2}\sin\frac{\beta}{2}$$

$$= \sinh\frac{0}{2}\cos\frac{\beta}{2} + j\cosh\frac{0}{2}\sin\frac{\beta}{2} = 0\times\cos\frac{\beta}{2} + j1\times\sin\frac{\beta}{2}$$

$$= j\sin\frac{\beta}{2}$$

$$(31b) \quad \sin\frac{\beta}{2} = \sqrt{\frac{m^2\omega^2/\omega_c^2}{1-(1-m^2)\omega^2/\omega_c^2}}$$

In the attenuation band $\gamma = \alpha + j\beta = \alpha + j\pi$. The attenuation α increases as frequency ω ranges from ω_c to infinity. The phase β has to be π to produce the imaginary result $j\cosh \alpha/2$.

$$(31c) \quad j\sqrt{\frac{m^2\omega^2/\omega_c^2}{1-(1-m^2)\omega^2/\omega_c^2}} = \sinh\frac{\gamma}{2} = \sinh\frac{\alpha}{2}\cos\frac{\beta}{2} + j\cosh\frac{\alpha}{2}\sin\frac{\beta}{2}$$

$$= \sinh\frac{\alpha}{2}\cos\frac{\pi}{2} + j\cosh\frac{\alpha}{2}\sin\frac{\pi}{2} = \sinh\frac{\alpha}{2}\times0 + j\cosh\frac{\alpha}{2}\times1$$

$$= j\cosh\frac{\alpha}{2}$$

$$(31d) \quad \cosh\frac{\alpha}{2} = \sqrt{\frac{m^2\omega^2/\omega_c^2}{1-(1-m^2)\omega^2/\omega_c^2}}$$

$z_{2m}=0$ at its resonant frequency $f_{2\infty}$, which produces equations for m and the theoretical infinite attenuation frequency ω_∞ (equation 32a).

$$(32) \quad z_{2m} = 0 = \frac{z_{2k}}{m} + z_{1k}\frac{1-m^2}{4m} \rightarrow 0 = \frac{1}{mp_\infty C_k} + p_\infty L_k\frac{1-m^2}{4m}$$

$$0 = 1 + p_\infty^2 L_k C_k\frac{1-m^2}{4} \rightarrow 0 = 1 + (1-m^2)\frac{j^2\omega_\infty^2}{4\omega_0^2} \rightarrow (1-m^2) = -\frac{4\omega_0^2}{j^2\omega_\infty^2}$$

$$(32a) \quad m = \sqrt{1-\frac{4\omega_0^2}{\omega_\infty^2}} = \sqrt{1-\frac{\omega_c^2}{\omega_\infty^2}} \rightarrow \omega_\infty^2 = \frac{\omega_c^2}{(1-m^2)} = \frac{4}{(1-m^2)L_k C_k}$$

Problem 205 Ref (29) Derive m derived π equation $Z_{1\pi m}^2$ for Figure 206

32

2.5 n Derived Filter Sections

Somebody discovered that image impedance is more nearly constant in the pass band when the m derived z_{1m} and z_{2m} are converted as follows.

$(33a)$ $\quad z_{2n} = \dfrac{z_{2k}}{n} = \dfrac{1}{npC_k}$

$(33b)$ $\quad \dfrac{1}{z_{1n}} = \dfrac{1}{nz_{1k}} + \dfrac{1}{z_{2k}}\dfrac{1-n^2}{4n} = \dfrac{1}{npL_k} + pC_k\dfrac{1-n^2}{4n} = \dfrac{4 + p^2 L_k C_k(1-n^2)}{4npL_k}$

Figure 207 n Derived Low Pass T and π sections

$(34a)$ $\qquad z_{I\,constant\,k} = z_{I\,n\,derived}$

$(34b)$ $\quad z_{ITn}^{\;2} = z_{1n} z_{2n} + \dfrac{z_{1n}^{\;2}}{4}$

$$= \dfrac{4npL_k}{4 + p^2 L_k C_k(1-n^2)}\dfrac{1}{npC_k} + \dfrac{1}{4}\left(\dfrac{4npL_k}{4 + p^2 L_k C_k(1-n^2)}\right)^2$$

$$= \dfrac{1}{4 + p^2 L_k C_k(1-n^2)}\,4\dfrac{L_k}{C_k}\left[1 + \dfrac{n^2 p^2 L_k C_k}{4 + p^2 L_k C_k(1-n^2)}\right]$$

$$= \dfrac{1}{4 + p^2 L_k C_k(1-n^2)}\,4\dfrac{L_k}{C_k}\left[\dfrac{4 + p^2 L_k C_k}{4 + p^2 L_k C_k(1-n^2)}\right]$$

$$z_{ITn} = \dfrac{1}{4}\dfrac{1}{1 + \frac{1}{4}p^2 L_k C_k(1-n^2)}\,2\sqrt{\dfrac{L_k}{C_k}} \times 2\sqrt{1 + \tfrac{1}{4}p^2 L_k C_k}$$

$(34c)$ $\quad z_{ITn} = \dfrac{1}{1 - \frac{\omega^2}{\omega_c^2}(1-n^2)}\sqrt{\dfrac{L_k}{C_k}\left(1 - \dfrac{\omega^2}{\omega_c^2}\right)}$

Problem 206 Plot z_{ITn} *when* $\dfrac{L_k}{C_k} = 1$ *and* $n = 0.6$ from $\omega = 0$ to ω_c.

2.6 Transfer Function Plots of Filter Sections

Low Pass Constant k and m components (Design page 42)

$(22f)\quad L_k = \dfrac{R}{\pi f_c} = \dfrac{600}{\pi 10^4} = 19.1 \times 10^{-3}$

$(22g)\quad C_k = \dfrac{1}{\pi f_c R} = \dfrac{1}{\pi 10^4 600} = 0.053 \times 10^{-6}$

$(27a)\quad L_{1m} = mL_k = 0.6 \times 19.1 \times 10^{-3} = 12.6 \times 10^{-3}$

$(30)\quad L_{2m} = L_k \dfrac{1-m^2}{4m} = 19.1 \times 10^{-3} \dfrac{1-0.36}{4 \times 0.6} = 5.09 \times 10^{-3}$

$\quad\quad C_{2m} = mC_k = 0.6 \times 0.053 \times 10^{-6} = 0.032 \times 10^{-6}$

$m^2 = 1 - \left(\dfrac{f_c}{f_\infty} \right)^2 \rightarrow \dfrac{f_c}{f_\infty} = \sqrt{1-m^2} = \sqrt{1-0.6^2} = \sqrt{0.64} = 0.8 \rightarrow f_\infty = 1.25 f_c$

Figure 205

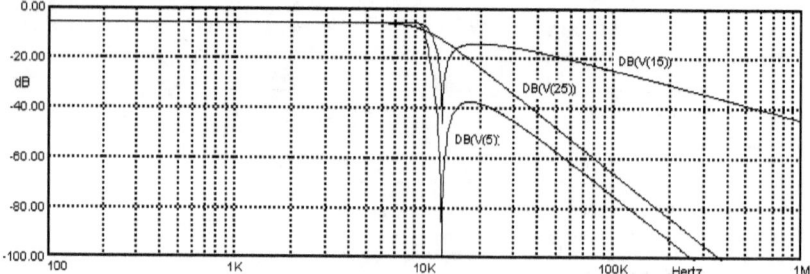

Figure 20511 Filter Transfer Functions – one k and one m section

Figure 20512 Filter Transfer Function –one k and one m section

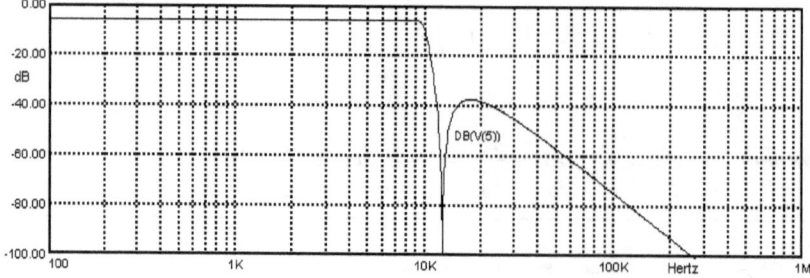

Fig2051.ckt Low pass filters

```
V1 1 0 AC 1 0
******
R21  1 22   600      ; constant k      fc=10KHz
L22k 22 24  9.55m    ; 0.5Lk
C22k 24 0   0.053u   ; Ck
L23k 24 25  9.55m    ; 0.5Lk
R22  25 0   600
******
R11  1 12   600      ; m derived      f∝=1.25×10KHz
L11m 12 13  6.3m     ; 0.5L1m
L12m 13 14  10.18m   ; 2L2m
C12m 14 0   0.016u   ; 0.5C2m
L13m 13 15  6.3m     ; 0.5L1m
R12  15 0   600
******
R1 1 2    600      ; cascade k and m
L1 2 3    10.18m   ; 2L2m
C1 3 0    0.016u   ; 0.5C2m
L2 2 7    6.3m     ; 0.5L1m

L3 7 4    9.55m    ; 0.5Lk
C2 4 0    0.053u   ; 0.5Ck
L4 4 8    9.55m    ; 0.5Lk

L5 8 5    6.3m     ; 0.5L1m
L6 5 6    10.18m   ; 2L2m
C4 6 0    0.016u   ; 0.5C2m
R2 5 0    600
******
*.AC DEC 201 100 1e+008
*.PLOT AC VDB(5) VDB(15) VDB(25) -100,0
.AC DEC 201 100 1e+006
.TEMP 27
.PLOT AC VDB(5) -100,0
.PRINT AC VDB(15) VDB(25)
.PRINT AC VDB(5)
.end
```

Analog Filter Design

Low Pass Add an m derived section (f_∞=30KHz)

$$\text{If } f_\infty = 3f_c \text{ then } m = \sqrt{1 - \left(\frac{f_c}{f_\infty}\right)^2} = \sqrt{1 - \left(\frac{10}{30}\right)^2} = 0.94281$$

m components

$$(22f) \quad L_k = \frac{R}{\pi f_c} = \frac{600}{\pi 10^4} = 19.1 \times 10^{-3}$$

$$(22g) \quad C_k = \frac{1}{\pi f_c R} = \frac{1}{\pi 10^4 600} = 0.053 \times 10^{-6}$$

$$(27a) \quad L_{1m} = mL_k = 0.94281 \times 19.1 \times 10^{-3} = 18 \times 10^{-3}$$

$$(30) \quad L_{2m} = L_k \frac{1 - m^2}{4m} = 19.1 \times 10^{-3} \frac{1 - 0.88889}{4 \times 0.94281} = 0.563 \times 10^{-3}$$

$$C_{2m} = mC_k = 0.94281 \times 0.053 \times 10^{-6} = 0.050 \times 10^{-6}$$

Figure 20521 Filter Transfer Function – Low Pass one k and two m sections

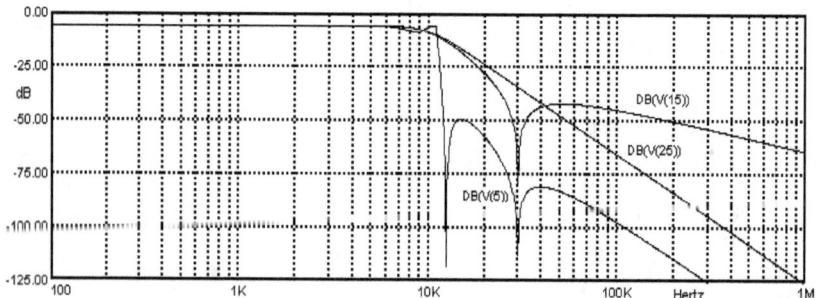

Figure 20522 Filter Transfer Function – Low Pass one k and two m sections

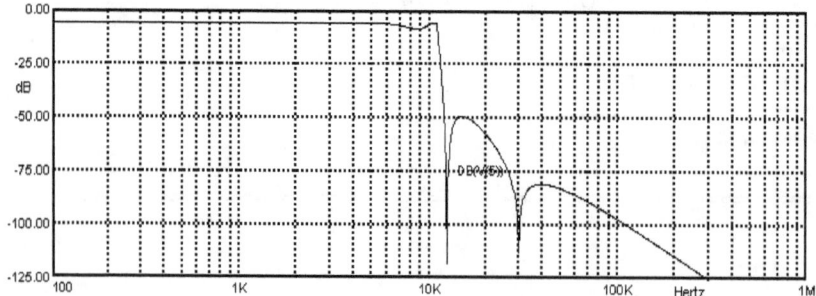

Fig2052.ckt Low pass filters

```
V1 1 0 AC 1 0
******
R21  1 22    600       ; constant k  10KHz
L22k 22 24   9.55m     ; 0.5Lk
C22k 24  0   0.053u    ; Ck
L23k 24 25   9.55m     ; 0.5Lk
R22  25 0    600
******
R11  1 12    600       ; m derived  3×10KHz
L11m 12 13   9m        ; 0.5L1m
L12m 13 14   0.563m    ; L2m
C2m  14  0   0.050u    ; C2m
L13m 13 15   9m        ; 0.5L1m
R12  15 0    600
******
R1 1 2    600       ; cascade k and m and m
L1 2 3    10.18m    ; 2L2m  m derived 1.25×10KHz
C1 3 0    0.016u    ; 0.5C2m

L2 2 4    9.55m     ; 0.5Lk constant k
C2 4 0    0.053u    ; Ck
L3 4 42   9.55m     ; 0.5Lk

L21m 42 43  9m       ; 0.5LIM  m derived 3×10KHz
L22m 43 44  0.563m   ; L2m
C22m 44 0   0.050u   ; C2m
L31m 43 5   9m       ; 0.5L1m

L4 5 6    10.18m    ; 2L2m  m derived 1.25×10KHz
C4 6 0    0.016u    ; 0.5C2m
R2 5 0    600
******
*.PLOT AC VDB(5) VDB(15) VDB(25) -125,0
.AC DEC 301 100 1e+006
.TEMP 27
.PLOT AC VDB(5) -125,0
.PRINT AC VDB(5) VDB(15) VDB(25)
.end
```

Analog Filter Design

High Pass Constant k and m components

$$(35a) \quad L_k = \frac{R}{4\pi f_c} = \frac{600}{4\pi 10^4} = 4.77 \times 10^{-3}$$

$$(35b) \quad C_k = \frac{1}{4\pi f_c R} = \frac{1}{4\pi 10^4 600} = 0.0133 \times 10^{-6}$$

$$(35c) \quad C_{1m} = \frac{C_k}{m} = \frac{0.0133 \times 10^{-6}}{0.6} = 0.02217 \times 10^{-6}$$

$$(35d) \quad C_{2m} = C_k \frac{4m}{1-m^2} = 0.0133 \times 10^{-6} \frac{4 \times 0.6}{1-0.36} = 0.0499 \times 10^{-6}$$

$$L_{2m} = \frac{L_k}{m} = \frac{4.77 \times 10^{-3}}{0.6} = 7.95 \times 10^{-3}$$

$$(35e) \quad m^2 = 1 - \left(\frac{f_\infty}{f_c}\right)^2 \rightarrow \frac{f_\infty}{f_c} = \sqrt{1-m^2} = \sqrt{1-0.6^2} = \sqrt{0.64} = 0.8 \rightarrow f_\infty = 0.8 f_c$$

Figure 218

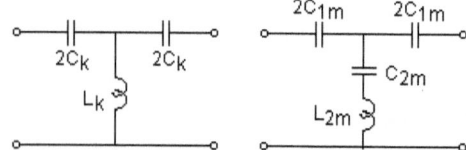

Figure 21811 Filter Transfer Function – High Pass one k and one m sections

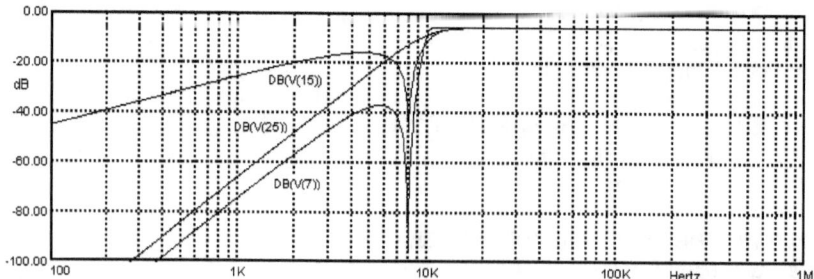

Figure 21812 Filter Transfer Function – High Pass one k and one m sections

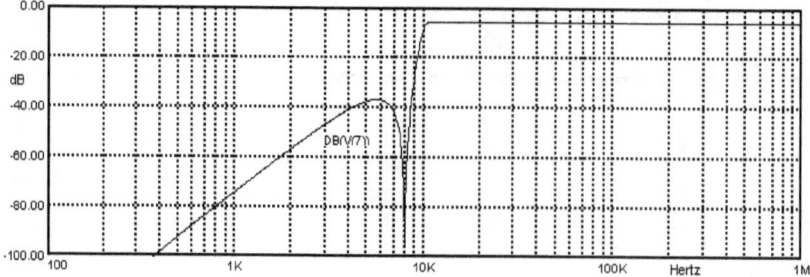

Fig2181.ckt High pass filters

```
V1 1 0 AC 1 0
******
R21  1 22    600         ; constant k 10KHz
C23k 22 24    0.0266u     ; 2Ck
L22k 24  0   4.77m        ; Lk
C21k 24 25    0.0266u     ; 2Ck
R22  25  0   600
******
R11  1 12    600         ; m derived 0.8×10KHz
C11m 12 13  0.04434u   ; 2C1m
L2m  14  0   7.95m        ; L2m
C2m  13 14   0.0499u      ; C2m
C12m 13 15  0.04434u   ; 2C1m
R12  15  0   600
R0   13  0   1e12
******
R1 1 2    600        ; cascade k and m
C1 2 3    0.0250u      ; C2m/2   m derived 0.8×10KHz
L1 3 0    15.9m        ; 2L2m
C2 2 4    0.04434u   ; 2C1m

C3 4 5    0.0266u      ; 2Ck    10KHz
L2 5 0    4.77m        ; Lk
C4 5 6    0.0266u      ; C2m

C5 6 7    0.04434u   ; 2C1m    m derived 0.8×10KHz
C6 7 8    0.0250u      ; C2m/2
L4 8 0    15.9m        ; 2L2m
R2 7 0    600
R00  4 0 1e12
R000 6 0 1e12
******
*.AC DEC 201 100 1e+008
*.PLOT AC VDB(7) VDB(15) VDB(25) -100,0
.AC DEC 201 100 1e+006
.TEMP 27
.PLOT AC VDB(7) -100,0
.PRINT AC VDB(15) VDB(25)
.PRINT AC VDB(7)
.end
```

Analog Filter Design

High Pass Add an m derived section ($f_\infty = f_c/3 = 3.333\text{KHz}$)

m components

$$(36a)\quad m^2 = 1 - \left(\frac{f_\infty}{f_c}\right)^2 = 1 - \left(\frac{1}{3}\right)^2 = 1 - \frac{1}{9} = 0.88889 \quad \rightarrow \quad m = 0.94281$$

$$(36a)\quad L_k = \frac{R}{4\pi f_c} = \frac{600}{4\pi 10^4} = 4.77 \times 10^{-3}$$

$$(36b)\quad C_k = \frac{1}{4\pi f_c R} = \frac{1}{4\pi 10^4 600} = 0.0133 \times 10^{-6}$$

$$(36c)\quad C_{1m} = \frac{C_k}{m} = \frac{0.0133 \times 10^{-6}}{0.94281} = 0.0141 \times 10^{-6}$$

$$(36d)\quad C_{2m} = C_k \frac{4m}{1-m^2} = 0.0133 \times 10^{-6} \frac{4 \times 0.94281}{1 - 0.8889} = 0.451 \times 10^{-6}$$

$$L_{2m} = \frac{L_k}{m} = \frac{4.77 \times 10^{-3}}{0.94281} = 5.06 \times 10^{-3}$$

Figure 21821 Filter Transfer Function – High Pass one k and two m sections

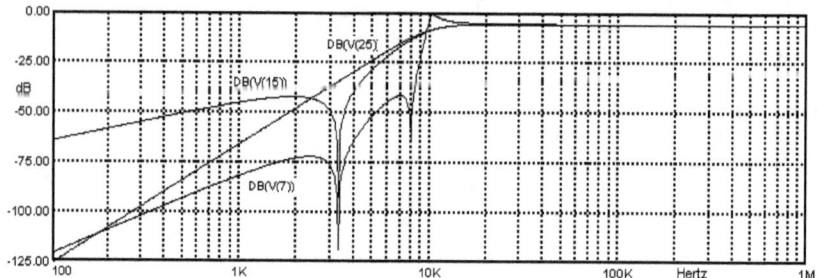

Figure 21821 Filter Transfer Function – High Pass one k and two m sections

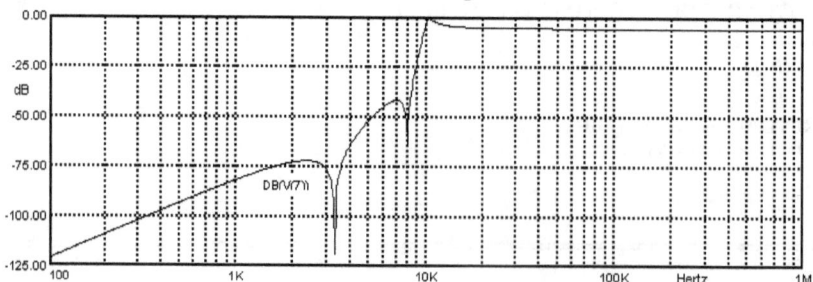

Fig2182.ckt high pass filters

```
V1 1 0 AC 1 0
******
R21  1 22    600          ; constant k 10KHz
C23k 22 24    0.0266u      ; 2Ck
L22k 24 0   4.77m       ; Lk
C21k 24 25    0.0266u      ; 2Ck
R22  25 0   600
******
R11  1 12    600          ; m derived 0.333×10KHz
C11m 12 13  0.0282u       ; 2C1m
L2m  14 0   5.06m       ; L2m
C2m  13 14  0.451u      ; C2m
C12m 13 15  0.0282u       ; 2C1m
R12  15 0   600
R4  13 0    1e12
******
R1 1 2    600          ; cascade k and m and m
C1 2 3    0.0250u       ; C2m/2  m derived 0.8×10KHz
L1 3 0    15.9m       ; 2L2m
C2 2 4    0.04434u      ; 2C1m

C3 4 5    0.0266u       ; 2Ck   constant k 10KHz
L2 5 0    4.77m       ; Lk
C4 5 6    0.0266u       ; 2Ck
Ra 4 0    1e12
Rb 6 0    1e12

C5 6 7    0.0282u       ; 2C1m   m derived 0.333×10KHz
C6 7 8    0.451u       ; C2m
L3 8 0    5.06m       ; L2m
C7 7 9    0.0282u       ; 2C1m

C8  9 10 04434u       ; 2C1m      m derived 0.8×10KHz
C9 10 11 0.0250u       ; C2m/2
L4 11 0   15.9m       ; 2L2m
R2 10 0  600
Rc 7 0   1e12
Rd 9 0   1e12
******
*.AC DEC 201 100 1e+008
*.PLOT AC VDB(7) -125,0
.AC DEC 201 100 1e+006
.TEMP 27
.PLOT AC VDB(7) VDB(15) VDB(25) -125,0
.PRINT AC VDB(7) VDB(15) VDB(25)
.end
```

Design– Low Pass

R = source and load resistors

$$m = \sqrt{1 - \left(\frac{f_c}{f_\infty}\right)^2}$$

Constant k section also known as the prototype.

$$m = 1 \quad f_\infty = \infty \quad L_k = \frac{R}{\pi f_c} \quad C_k = \frac{1}{\pi f_c R}$$

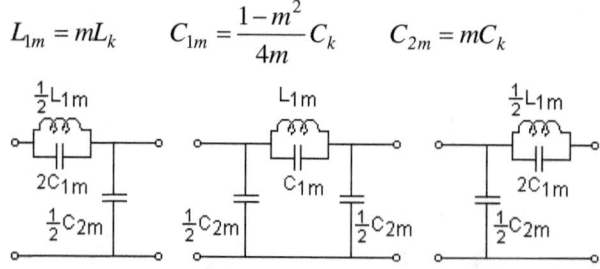

m Derived T section

$$L_{1m} = mL_k \qquad L_{2m} = \frac{1-m^2}{4m} L_k \qquad C_{2m} = mC_k$$

m Derived π section

$$L_{1m} = mL_k \qquad C_{1m} = \frac{1-m^2}{4m} C_k \qquad C_{2m} = mC_k$$

Problem 207 Design low pass filter, 1 k section, 2 m derived sections, R=50Ω, f_c=1MHz, f_∞= 1.25f_c, 2f_c. Write Spice program. Plot filter transfer function.

Design–High Pass

R = source and load resistors

$$m = \sqrt{1 - \left(\frac{f_\infty}{f_c}\right)^2}$$

Constant k section also known as the prototype.

$$m = 1 \quad f_\infty = 0 \quad L_k = \frac{R}{4\pi f_c} \quad C_k = \frac{1}{4\pi f_c R}$$

m Derived T section

$$C_{1m} = \frac{C_k}{m} \quad C_{2m} = \frac{4m}{1 - m^2} C_k \quad L_{2m} = \frac{L_k}{m}$$

m Derived π section

$$L_{1m} = \frac{4m}{1 - m^2} L_k \quad C_{1m} = \frac{C_k}{m} \quad L_{2m} = \frac{L_k}{m}$$

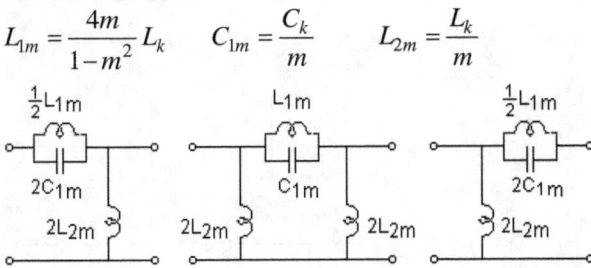

Problem 208 Design high pass filter, 1 k section, 2 m derived sections, R=50Ω, f_c=10MHz, f_∞= 0.8f_c, 0.5f_c. Write Spice program. Plot filter transfer function.

Design–Band Pass

R = source and load resistors

$$m^2 = 1 - \left(\frac{f_\infty [f_2 - f_1]}{f_\infty^2 - f_2 f_1} \right)^2 \qquad f_\infty = f_{1\infty} \ or \ f_{2\infty} \qquad f_{1\infty} f_{2\infty} = f_1 f_2$$

Constant k section also known as the prototype.

$$m = 1 \qquad f_{1\infty} = 0 \qquad f_{2\infty} = \infty$$

$$L_{1bk} = \frac{R}{\pi(f_2 - f_1)} \qquad C_{1bk} = \frac{f_2 - f_1}{4\pi f_1 f_2 R} \qquad L_{2bk} = \frac{R(f_2 - f_1)}{4\pi f_1 f_2} \qquad C_{2bk} = \frac{1}{\pi(f_2 - f_1)R}$$

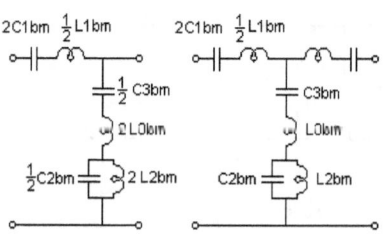

m Derived T section

$$L_{1bm} = mL_{1bk} \qquad C_{1bm} = \frac{C_{1bk}}{m}$$

$$L_{2bm} = \frac{L_{2bk}}{m} \qquad C_{2bm} = mC_{2bk}$$

$$L_{3bm} = \frac{1-m^2}{4m} L_{1bk} \qquad C_{3bm} = \frac{4m}{1-m^2} C_{1bk}$$

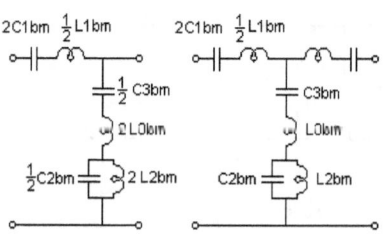

m Derived π section

$$L_{1bm} = mL_{1bk} \qquad C_{1bm} = \frac{C_{1bk}}{m}$$

$$L_{2bm} = \frac{L_{2bk}}{m} \qquad C_{2bm} = mC_{2bk}$$

$$C_{4bm} = \frac{1-m^2}{4m} C_{2bk} \qquad L_{4bm} = \frac{4m}{1-m^2} L_{2bk}$$

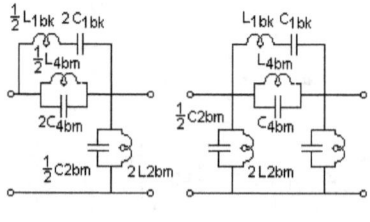

Problem 209 Design band pass filter, 1 k section, 1 m derived section, R=600Ω, f_1=100KHz, f_2=110KHz, f_∞= 1.25f_2. Write Spice program. Plot filter transfer function.

3 Modern Filter Design

Chapter 3 and Chapter 4 show how to design modern LC analog filters. We start by showing how filters are specified. The specifications as written are not physically realizable. Approximations to the specifications are physically realizable by a circuit with real components. We show how to implement the Butterworth, Bessel, Chebyshev, and Inverse Chebyshev approximations, which are physically realizable rational functions of p. The approximations produce a ratio of two polynomials $N(p)/D(p)$, which we write as $T(p)$ where $T(p)$ is the output/input transfer function of a low pass filter. If $T(p)$ has no finite zeros, then $N(p)$ is just a number such as 1.

Analog filter design has three major steps
  Specify the pass and attenuation frequency response bands.
  Find a transfer function approximating the specification.
  Design a circuit that implements the transfer function.

In Chapter 4 we implement the last step, designing circuits with real components that implement $T(p)$. In addition, frequency transformations convert $T(p)$ into high pass, band pass, and band reject configurations.

The Butterworth, Bessel, Chebyshev, and Inverse Chebyshev approximations are used most of the time. Almost all applications can be fulfilled with one of these. Know that other approximations are available.

Butterworth has a flat response in the pass band, and a $-20n$ dB/decade cutoff, where integer n is the order of the filter. The Butterworth filter is straightforward to understand. Its flat response makes it suitable for audio systems, whereas the Chebyshev pass band fluctuations make it unsuitable for those audio systems. Inverse Chebyshev has a flat response pass band.

The Bessel filter has a constant propagation delay for all frequencies. This means a pulse driving the input of a Bessel filter produces an output pulse with essentially no overshoot or undershoot distortions. Other filters do not delay the harmonics by the same amount of time, which is why their digital output waveforms are distorted.

3.1 Filter Specifications

All filter specifications acknowledge the existence of the approximation problem by the presence of shaded bands for the magnitude of the transfer function T, and transition bands between pass and stop bands. The magnitude of the transfer function T is specified to fall within the shaded bands, which is the goal of a filter design.

For example, specifying a low pass filter -3dB cutoff frequency ω_1 (Figure 301) is straightforward, however realizing the specified ω_2 in combination with attenuation β can be difficult when cost considerations are taken into account. *Many aspects of filter design are an art form in the sense there is no theory that guarantees one can achieve the goal on the first try.*

Filters are referred to as attenuation equalizers when magnitude is specified as a function of frequency. Filters are referred to as phase equalizers[5] when phase is specified as a function of frequency. *This text is about attenuation equalizers.*

Figure 301 Low Pass Filter Specification

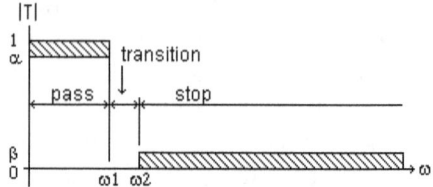

Figure 302 High Pass Filter Specification

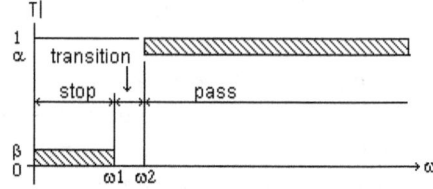

Figure 303 Band Pass Filter Specification

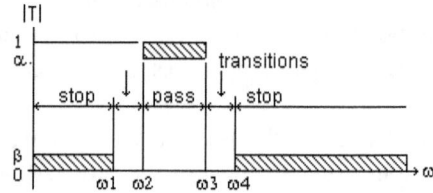

Figure 304 Band Reject Filter Specification

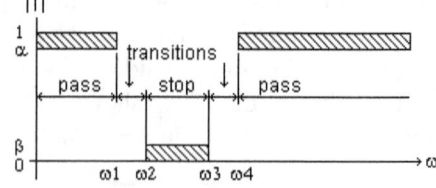

---

[5] H. W. Bode, Network analysis and Feedback Design, 1945

3.2 Butterworth Approximation to a Low Pass Filter

Butterworth[6] discovered the magnitude function $M_{bun}(\omega)$ for a *maximally flat* approximation to a low pass filter specification by setting the first 2n−1 derivatives of a general transfer function to zero at $\omega=0$.

(1) $\quad M_{bun}(\omega) = \dfrac{1}{\sqrt{1+\omega^{2n}}}$ $\qquad$ *and* $\quad M_{bun}(1) = \dfrac{1}{\sqrt{2}}$ $\quad$ (−3dB)

This approximation is said to be maximally flat, because the first 2n−1 derivatives of $M_{bun}(\omega)$ equal 0 at $\omega=0$. This is clear from the power series expansion of $M_{bun}(\omega)$.

(2) $\quad M_{bun}(\omega) = 1 - \tfrac{1}{2}\omega^{2n} + \tfrac{3}{8}\omega^{4n} - \tfrac{5}{16}\omega^{6n} + \cdots$

The asymptote of the magnitude is ω^{-n} when $\omega \gg 1$. That is to say the asymptotic slope is −6n dB/octave or −20n dB/decade.

(3) $\quad 20\log M_{bun}(\omega) = 20\log\omega^{-n} = -20n\log\omega$ $\quad$ (dB) $\qquad$ ($\omega > 1$)

Deriving the Transfer function $T_{bun}(p)$

(4) $\quad |z|^2 = z\bar{z}$

Figure 305   n=3

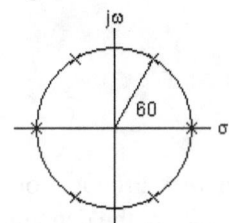

if $\left|T_{bun}(j\omega)\right| = M_{bun}(\omega)$ $\quad$ *then*

(5) $\quad T_{bun}(j\omega)T_{bun}(-j\omega) = M_{bun}^{2}(\omega) = \dfrac{1}{1+\omega^{2n}}$

To find T we recognize that

when $p = j\omega,$ $\quad p^2 = -\omega^2$ $\quad$ *so that*

(6) $\quad T_{bun}(p)T_{bun}(-p) = \dfrac{1}{1+(-p^2)^n}$

The poles p_k of $T_{bun}(p)$ and $T_{bun}(-p)$ are zeros of the denominator. The poles p_k are on the unit circle (Figure 305), because the magnitude of every p_k equals 1 (equation 9). The poles in the left hand plane are assigned to $T_{bun}(p)$, and the poles in the right hand plane are assigned to $T_{bun}(-p)$.

(7) $\quad 1+(-p_k^{2})^n = 0$ $\quad \rightarrow \quad$ $(-1)^n p_k^{2n} = -1$ $\quad \rightarrow \quad$ $(e^{j\pi})^n p_k^{2n} = e^{j(\pi+2\pi k)}$

(8) $\quad p_k^{2n} = e^{j(\pi+2\pi k - \pi n)}$ $\qquad k = 0,\ 1,\ 2,\ 3,\dots$

(9) $\quad p_k = \sigma_k + j\omega_k = e^{j\theta_{nk}} = \cos\theta_{nk} + j\sin\theta_{nk}$ $\quad$ *where* $\theta_{nk} = \dfrac{\pi}{2n} + \dfrac{\pi k}{n} - \dfrac{\pi}{2}$

[6] S. Butterworth , "On the Theory of Filter Amplifiers", Wireless Engineer, vol. 7, 1930, pp. 536–541

Pole locations The zeros $p_k = \sigma_k + j\omega_k$ of a polynomial (equation 7) can be calculated to any required accuracy from the equations 10 θ_{nk} angles. This is how Table 303 is constructed.

$(10a)\quad \sigma_k = \cos\theta_{nk} = \cos\left(\frac{\pi}{2n} + \frac{\pi k}{n} - \frac{\pi}{2}\right) \qquad k = 0,\ 1,\ 2, \cdots,\ n-1$

$(10b)\quad \omega_k = \sin\theta_{nk} = \sin\left(\frac{\pi}{2n} + \frac{\pi k}{n} - \frac{\pi}{2}\right)$

$(11a)\quad \theta_{nk} = \left(\frac{\pi}{2n} + \frac{\pi k}{n} - \frac{\pi}{2}\right) \qquad k = 0,\ 1,\ 2, \cdots,\ n-1$

$(11b)\quad \theta_{2k} = \left(\frac{\pi}{4} + \frac{\pi k}{2} - \frac{2\pi}{4}\right) \quad \theta_{3k} = \left(\frac{\pi}{6} + \frac{\pi k}{3} - \frac{3\pi}{6}\right) \quad \theta_{4k} = \left(\frac{\pi}{8} + \frac{\pi k}{4} - \frac{4\pi}{8}\right)$

$\qquad\quad \theta_{5k} = \left(\frac{\pi}{10} + \frac{\pi k}{5} - \frac{5\pi}{10}\right) \quad \theta_{6k} = \left(\frac{\pi}{12} + \frac{\pi k}{6} - \frac{6\pi}{12}\right) \quad \theta_{7k} = \left(\frac{\pi}{14} + \frac{\pi k}{7} - \frac{7\pi}{14}\right)$

$\qquad\quad \theta_{8k} = \left(\frac{\pi}{16} + \frac{\pi k}{8} - \frac{8\pi}{16}\right) \qquad$ *Add π for θ in left half plane*

k	θ_{2k}	θ_{3k}	θ_{4k}	θ_{5k}	θ_{6k}	θ_{7k}	θ_{8k}
0	$-\frac{\pi}{4}$	$-\frac{2\pi}{6}$	$-\frac{3\pi}{8}$	$-\frac{4\pi}{10}$	$-\frac{5\pi}{12}$	$-\frac{6\pi}{14}$	$-\frac{7\pi}{16}$
1	$+\frac{\pi}{4}$	0	$-\frac{\pi}{8}$	$-\frac{2\pi}{10}$	$-\frac{3\pi}{12}$	$-\frac{4\pi}{14}$	$-\frac{5\pi}{16}$
2		$+\frac{2\pi}{6}$	$+\frac{\pi}{8}$	0	$-\frac{\pi}{12}$	$-\frac{2\pi}{14}$	$-\frac{3\pi}{16}$
3			$+\frac{3\pi}{8}$	$+\frac{2\pi}{10}$	$+\frac{\pi}{12}$	0	$-\frac{\pi}{16}$
4				$+\frac{4\pi}{10}$	$+\frac{3\pi}{12}$	$+\frac{2\pi}{14}$	$+\frac{\pi}{16}$
5					$+\frac{5\pi}{12}$	$+\frac{4\pi}{14}$	$+\frac{3\pi}{16}$
6						$+\frac{6\pi}{14}$	$+\frac{5\pi}{16}$
7							$+\frac{7\pi}{16}$

Factors Products of zeros produce the polynomials $B_{un}(p)$ in Table 302, which are the denominators of Butterworth transfer functions $T_{bun}(p)$.

Polynomials Products of factors create polynomials $B_{un}(p)$ of Butterworth transfer functions $T_{bun}(p)$ (Table 301).

Example From Tables 301, 302, 303 we construct transfer function $T_{bu5}(p)$ in its three forms.

$(12a)\quad \dfrac{1}{T_{bu5}(p)} = B_{u5}(p) = (p + 0.8090 + j0.5878)(p + 0.8090 - j0.5878)$

$\qquad\qquad\qquad (p + 0.3090 + j0.9511)(p + 0.3090 - j0.9511)(p + 1)$

$(12b)\quad T_{bu5}(p) = \dfrac{1}{B_{u5}(p)} = \dfrac{1}{(p^2 + 0.6180p + 1)(p^2 + 1.6180p + 1)(p + 1)}$

$(12c)\quad T_{bu5}(p) = \dfrac{1}{B_{u5}(p)} = \dfrac{1}{p^5 + 3.236p^4 + 5.236p^3 + 5.236p^2 + 3.236p + 1}$

Table 301 Butterworth Polynomials $B_{un}(p)$

n	$B_{bun}(p)$

1 $p+1$

2 $p^2 + \sqrt{2}p + 1$

3 $p^3 + 2p^2 + 2p + 1$

4 $p^4 + 2.613p^3 + 3.414p^2 + 2.613p + 1$

5 $p^5 + 3.236p^4 + 5.236p^3 + 5.236p^2 + 3.236p + 1$

6 $p^6 + 3.864p^5 + 7.464p^4 + 9.141p^3 + 7.464p^2 + 3.864p + 1$

7 $p^7 + 4.494p^6 + 10.098p^5 + 14.592p^4 + 14.592p^3 + 10.098p^2 + 4.494p + 1$

8 $p^8 + 5.126p^7 + 13.137p^6 + 21.846p^5 + 25.688p^4 + 21.846p^3$
$$+ 13.137p^2 + 5.126p + 1$$

n *Table 302 Factors of the Butterworth $B_{un}(p)$ polynomials*

1 $p+1$

2 $p^2 + \sqrt{2}p + 1$

3 $(p^2 + p + 1)(p + 1)$

4 $(p^2 + 0.7654p + 1)(p^2 + 1.8478p + 1)$

5 $(p^2 + 0.6180p + 1)(p^2 + 1.6180p + 1)(p + 1)$

6 $(p^2 + 0.5176p + 1)(p^2 + \sqrt{2}p + 1)(p^2 + 1.9318p + 1)$

7 $(p^2 + 0.4450p + 1)(p^2 + 1.2470p + 1)(p^2 + 1.8020p + 1)(p + 1)$

8 $(p^2 + 0.3902p + 1)(p^2 + 1.1112p + 1)(p^2 + 1.6630p + 1)(p^2 + 1.9616p + 1)$

Table 303 Zeros of the $B_{un}(p)$ polynomials

order	zeros		order	zeros
2	$-0.7071 \pm j0.7071$		7	$-0.9010 \pm j0.4339$
3	$-0.5000 \pm j0.8660$			$-0.6235 \pm j0.7818$
	$-1.0000 \pm j0$			$-0.2225 \pm j0.9749$
4	$-0.9239 \pm j0.3827$			$-1.0000 \pm j0$
	$-0.3827 \pm j0.9239$		8	$-0.9808 \pm j0.1951$
5	$-0.8090 \pm j0.5878$			$-0.8315 \pm j0.5556$
	$-0.3090 \pm j0.9511$			$-0.5556 \pm j0.8315$
	$-1.0000 \pm j0$			$-0.1951 \pm j0.9808$
6	$-0.9659 \pm j0.2588$			
	$-0.7071 \pm j0.7071$			
	$-0.2588 \pm j0.9659$			

Analog Filter Design

Example: Verify Butterworth $T_{bu3}(p)$ and $M_{bu6}(p)$

The zeros of the polynomials are the poles of $T_{bun}(p)$. The magnitude of each pole $|p_k|=1$. The angle of each pole θ_{3k} is calculated from equation 11. Add π to get the left hand plane angles when k=3, 4, 5. The poles in the left hand half plane are assigned to $T_{bu3}(p)$. The poles in the right hand half plane, k=0, 1, 2, are assigned to $T_{bu3}(-p)$.

$$p_k = e^{j\theta_{3k}} \qquad \theta_{3k} = \tfrac{\pi}{6} + \tfrac{\pi k}{3} - \tfrac{\pi}{2} = +\tfrac{\pi k}{3} - \tfrac{\pi}{3} \qquad k = 0,\ 1,\ 2,\ 3,\ 4,\ 5$$

$$\theta_{30} = -\tfrac{\pi}{3} \quad \theta_{31} = 0 \quad \theta_{32} = \tfrac{\pi}{3} \quad \theta_{33} = \tfrac{2\pi}{3} \quad \theta_{34} = \pi \quad \theta_{35} = \tfrac{4\pi}{3}$$

$$\theta_{30} = -60° \quad \theta_{31} = 0° \quad \theta_{32} = 60° \quad \theta_{33} = 120° \quad \theta_{34} = 180° \quad \theta_{35} = 240°$$

$$\frac{1}{T_{bu3}(p)} = B_{u3}(p) = (p - p_3)(p - p_4)(p - p_5)$$

$$= (p - e^{j\frac{2\pi}{3}})(p - e^{j\frac{3\pi}{3}})(p - e^{j\frac{4\pi}{3}}) = \left(p + \tfrac{1}{2} - j\tfrac{\sqrt{3}}{2}\right)(p+1)\left(p + \tfrac{1}{2} + j\tfrac{\sqrt{3}}{2}\right)$$

$$= (p+1)\left[\left(p + \tfrac{1}{2}\right)^2 + \left(\tfrac{\sqrt{3}}{2}\right)^2\right] = (p+1)(p^2 + p + 1)$$

$$= p^3 + 2p^2 + 2p + 1 \quad as\ written\ in\ Table\ 301\ n = 3$$

$$\frac{1}{T_{bu3}(-p)} = B_{u3}(p) = (-p + p_0)(-p + p_1)(-p + p_2)$$

$$= -(p - e^{-j\frac{\pi}{3}})(p - e^{j0})(p - e^{j\frac{\pi}{3}}) = -\left(p - \tfrac{1}{2} + j\tfrac{\sqrt{3}}{2}\right)(p-1)\left(p - \tfrac{1}{2} - j\tfrac{\sqrt{3}}{2}\right)$$

$$= -(p-1)\left[\left(p - \tfrac{1}{2}\right)^2 + \left(\tfrac{\sqrt{3}}{2}\right)^2\right] = -(p-1)(p^2 - p + 1)$$

$$= -p^3 + 2p^2 - 2p + 1$$

$$T_{bu3}(p)T_{bu3}(-p) = \frac{1}{p^3 + 2p^2 + 2p + 1} \times \frac{1}{-p^3 + 2p^2 - 2p + 1} = \frac{1}{1 - p^6}$$

check $(-p^3 + 2p^2 - 2p + 1)(p^3 + 2p^2 + 2p + 1) = 1 - p^6 = 1/M_{bu6}^2(p)$ qed

---

Problem 301 Derive $T_{bu2}(p)$ and $T_{bu2}(-p)$ from $M_{bu2}(\omega)$ (Table 301 n=2).
Problem 302 Derive $T_{bu6}(p)$ and $T_{bu6}(-p)$ from $M_{bu6}(\omega)$ (Table 301 n=6).

3.3 Bessel Approximation to a Low Pass Filter

The Laplace transform of a time delay τ (translation in time) shows that a waveform is not distorted by the time delay. The waveform f(t–τ) is identical to the waveform f(t). The *transmission is distortionless*.

$$If \quad F(p) = \int_0^\infty f(t)u(t)e^{-pt}dt, \quad then \quad when \quad F_\tau(p) = \int_0^\infty f(t-\tau)u(t-\tau)e^{-pt}dt$$

and substitute $\delta = t - \tau$

$$F_\tau(p) = \int_0^\infty f(\delta)u(\delta)e^{-p(\tau+\delta)}d\tau = e^{-p\tau}\int_0^\infty f(\delta)u(\delta)e^{-p\delta}d\tau = e^{-p\tau}F(p)$$

(13) *This means if* $f(t)u(t) \Leftrightarrow F(p)$ *then* $f(t-\tau)u(t-\tau) \Leftrightarrow e^{-p\tau}F(p)$

This specific Laplace transform offers formal proof that the necessary and sufficient condition for distortionless delay τ is the transcendental transfer function exp(j$\omega\tau$), which has a *linear phase* image. In other words distortionless delay results when the transfer function phase is *linear* in ω.

(14) *In the frequency domain* $p = j\omega$, *so that* $\left. e^{-p\tau}\right|_{p=j\omega} = e^{-j\omega\tau} \equiv e^{-j\theta}$

(15) $\tau = \dfrac{\theta}{\omega} = \dfrac{\theta}{2\pi f}$

When each harmonic in a waveform is delayed by the same amount of time τ (and, in some circuits, attenuated by the same amount k maintaining constant relative amplitude), the waveform is NOT distorted.

(16) $\theta(\omega) = \omega\tau,$ *delay* $\tau = \dfrac{\theta(\omega)}{\omega}$ *and group delay* $= -\dfrac{d\theta(\omega)}{d\omega}$

Networks with linear phase and constant amplitude transmission are desirable in digital networks transmitting pulses (e.g. 1's and 0's). A long time ago[7] Bessel polynomials B(p) were determined to approximate linear phase very well in the sense that overshoot of pulse "edges" was reduced by greater than a factor of ten when compared to Butterworth and Chebyshev polynomials.

---

[7] Thomson, W. E., "*Network with Maximally flat delay*", Wireless Eng., vol 29, Oct 1952, p256-263

There is a recursion formula for Bessel polynomials $B_{en}(p)$.

(17a) $B_{e0}(p) = 1$ (17b) $B_{e1}(p) = p+1$

(17c) $B_{en+1}(p) = (2n+1)B_{en}(p) + p^2 B_{en-1}(p)$

(18a) $B_{e0}(p) = 1$ (18b) $B_{e1}(p) = p+1$

(18c) $B_{e2}(p) = 3B_{e1}(p) + p^2 B_{e0}(p) = 3(p+1) + p^2(1) = p^2 + 3p + 3$

(18d) $B_{e3}(p) = 5B_{e2}(p) + p^2 B_{e1}(p) = 5(p^2 + 3p + 3) + p^2(p+1)$

$$= p^3 + 6p^2 + 15p + 15$$

(18e) $B_{e4}(p) = 7B_{e3}(p) + p^2 B_{e2}(p) = 7(p^3 + 6p^2 + 15p + 15) + p^2(p^2 + 3p + 3)$

$$= p^4 + 10p^3 + 45p^2 + 105p + 105 \quad and \ so \ forth$$

Transfer function The Bessel transfer function is simply

(19) $T_{ben}(p) = \dfrac{B_{en}(0)}{B_{en}(p)}$ so that $T_{ben}(0) = 1$

---

n	Table 304 Bessel Polynomials $B_{en}(p)$
0	1
1	$p+1$
2	$p^2 + 3p + 3$
3	$p^3 + 6p^2 + 15p + 15$
4	$p^4 + 10p^3 + 45p^2 + 105p + 105$
5	$p^5 + 15p^4 + 105p^3 + 420p^2 + 945p + 945$
6	$p^6 + 21p^5 + 210p^4 + 1260p^3 + 4725p^2 + 10395p + 10,395$

Table 305 Bessel polynomials – zeros are found by using Newton's method. However the zeros are not used here.

n	zeros	n	zeros
5	-3.6467386	1	$-1.0000000 \pm j0$
	$-3.3519564 \pm j1.7426614$	2	$-1.5000000 \pm j0.8660254$
	$-2.3246743 \pm j3.5710229$	3	$-2.3221854 \pm j0$
	$-4.2483594 \pm j0.8675097$		$-1.8389073 \pm j1.7543810$
6	$-3.7357084 \pm j2.6262723$	4	$-2.8962106 \pm j0.8672341$
	$-2.5159322 \pm j4.4926730$		$-2.1037894 \pm j2.6574180$

3.4 Chebyshev Approximation to a Low Pass Filter

Chebyshev (a.k.a. Tchebycheff) created the polynomials $C_n(\omega)$, which can be applied to filter design. The $C_n(\omega)$ are the *Chebyshev functions of the first kind*. Here is the definition of $C_n(\omega)$.

(20a) $\quad C_n(\omega) = \cos(n\cos^{-1}\omega) \quad |\omega| \leq 1$

(20b) $\quad C_n(\omega) = \cosh(n\cosh^{-1}\omega) \quad |\omega| \geq 1$

The Chebyshev filter magnitude function is $M_{cn}(\omega)$.

(21) $\quad M_{cn}(\omega) = \dfrac{1}{\sqrt{1 + \varepsilon^2 C_n{}^2(\omega)}} \quad$ *ω is a normalized variable ω/ω_{cutoff}*

Deriving the Transfer function $T_{cn}(p)$

(22) $\quad |z|^2 = z\bar{z}$

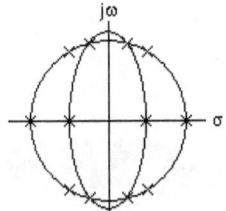

Figure 306 n=3

$\quad$ *if* $\ |T_{cn}(j\omega)| = M_{cn}(\omega) \quad$ *then*

(23) $\quad T_{cn}(j\omega)T_{cn}(-j\omega) = M_{cn}{}^2(\omega) = \dfrac{1}{1+\varepsilon^2 C_n{}^2(\omega)}$

To find $T_{cn}(p)$ recognize that
when $p = j\omega$, $\omega = -jp$ so that

(24) $\quad T_{cn}(p)T_{cn}(-p) = \dfrac{1}{1+\varepsilon^2 C_n{}^2(-jp)} = \dfrac{1}{D_n(p)D_n(-p)}$

Observe that $T_{cn}(p)$ *is* the transfer function. $T_{cn}(p)$ equals one over a polynomial we refer to as $D_n(p)$ that is *different* from $C_n(\omega)$.

Find poles of $T_{cn}(p)$ and $T_{cn}(-p)$ The poles of $T_{cn}(p)$ and $T_{cn}(-p)$ are the zeros of the denominator (24, 25a).

(25a) $\quad 1+\varepsilon^2 C_n{}^2(-jp) = 0 \ \rightarrow\ \varepsilon^2 C_n{}^2(-jp) = -1 = j^2 \ \rightarrow\ \varepsilon C_n(-jp) = j$

(25b) $\quad C_n(-jp) = 0 \pm j\dfrac{1}{\varepsilon} \quad \omega < 1$

(26) $\quad$ *since* $C_n(\omega) = \cos(n\cos^{-1}\omega) \qquad C_n(-jp) = \cos(n\cos^{-1}(-jp))$

The goal is the equations for the zeros $p_k = \sigma_k + j\omega_k$ of the denominator $D_n(p)$. Complex frequency p appears in equation 24. Extract p from arc cos $-jp$ (27b) and simplify (27c). Now values for u and v are required.

53

Analog Filter Design

(27a)   let $w = u + jv = \cos^{-1}(-jp)$ so that $\cos w = -jp$

(27b)   then $p = j\cos w = j\cos(u + jv)$

(27c) $p = j\cos u \cos jv - j\sin u \sin jv$

$p = j\cos u \cosh v - j\sin u(j\sinh v)$

$p = j\cos u \cosh v + \sin u \sinh v$

$p = \sigma + j\omega = \sin u \sinh v + j\cos u \cosh v$

After some preliminary tries equation 25b processed as equations 28b produced values for nu and nv (28c, 28d).

(28a)   let $w = u + jv = \cos^{-1}(-jp)$ so that $\cos nw = 0 \pm j\dfrac{1}{\varepsilon} = C_n(-jp)$

(28b) $0 \pm j\dfrac{1}{\varepsilon} = \cos nw = \cos(nu + jnv)$

$0 \pm j\dfrac{1}{\varepsilon} = \cos nu \cos jnv - \sin nu \sin jnv = \cos nu \cosh nv - j\sin nu \sinh nv$

(28c) $0 = \cos nu \cosh nv \;\rightarrow\; nu = \dfrac{\pi}{2}(2k-1) \quad k = 1, 2, ..., n$

since $\sin nu = \sin\dfrac{\pi}{2}(2k-1) = \pm 1$

(28d) $\pm\dfrac{1}{\varepsilon} = -\sin nu \sinh nv \;\rightarrow\; \pm\dfrac{1}{\varepsilon} = \pm 1\sinh nv \;\rightarrow\; nv = \sinh^{-1}\dfrac{1}{\varepsilon}$

Calculate ε given the dB in the pass band ripple.

passband ripple $r = \sqrt{1+\varepsilon^2} \;\rightarrow\; r_{dB} = 20\log r = 10\log(1+\varepsilon^2) \;\rightarrow\; \varepsilon = \sqrt{10^{\frac{r_{dB}}{10}} - 1}$

Substitute u and v values into 27c to get equations for the zeros (poles of T(p)).

(29a)   from 28c $u = \dfrac{\pi}{2n}(2k-1)$ and   from 28d $v = \dfrac{1}{n}\sinh^{-1}\dfrac{1}{\varepsilon}$ $k = 1, 2, ..., n$
(29b) $p_k = \sigma_k + j\omega_k = -\sinh v \sin\left(\dfrac{\pi}{2n}(2k-1)\right) + j\cosh v \cos\left(\dfrac{\pi}{2n}(2k-1)\right)$

Show that the zeros lie on an ellipse. From 27c we form the following.

(30a) $\sin^2\theta = \dfrac{\sigma_k^2}{\sinh^2 v}$ and $\cos^2\theta = \dfrac{\omega_k^2}{\cosh^2 v}$ $\rightarrow$ $\theta = \dfrac{\pi}{2n}(1+2k)$

(30b) $\sin^2\theta + \cos^2\theta = \dfrac{\sigma_k^2}{\sinh^2 v} + \dfrac{\omega_k^2}{\cosh^2 v} = 1$ $\rightarrow$ $v = \dfrac{1}{n}\sinh^{-1}\dfrac{1}{\varepsilon}$

Example: Chebyshev $T_{c3}(p)$ for n=3, 1dB ripple (ε=0.50885)

passband ripple $r = 1dB$ $\rightarrow$ $\varepsilon = \sqrt{10^{\frac{rdB}{10}} - 1} = \sqrt{10^{\frac{1}{10}} - 1} = 0.50885$

$v = \frac{1}{n}\sinh^{-1}\frac{1}{\varepsilon} = \frac{1}{3}\sinh^{-1}\frac{1}{0.50885} = 0.47599,$ $\sinh v = 0.49417,$ $\cosh v = 1.11544$

(29b) $p_k = \sigma_k \pm j\omega_k = -\sinh v \times \sin\left(\frac{\pi}{2n}(2k-1)\right) \pm j\cosh v \times \cos\left(\frac{\pi}{2n}(2k-1)\right)$

Chebyshev poles $p_{ck} = \sigma_{ck} + j\omega_{ck}$ $\quad n = 3 \quad k = 1, 2, 3$

$\sigma_{ck=1} = \sinh v \cdot \sin 1\frac{\pi}{6} = 0.24709$ $\quad \sigma_{ck=2} = \sinh v \cdot \sin 3\frac{\pi}{6} = 0.49417$

$\sigma_{ck=3} = \sinh v \cdot \sin 5\frac{\pi}{6} = 0.24709$ $\quad \omega_{ck=1} = \cosh v \cdot \cos 1\frac{\pi}{6} = 0.96600$

$\omega_{ck=2} = \cosh v \cdot \cos 3\frac{\pi}{6} = 0$ $\quad \omega_{ck=3} = \cosh v \cdot \cos 5\frac{\pi}{6} = 0.96600$

Form the poles.

$p_{c1}, p_{c3} = -0.24709 \mp j0.96600$ $\quad p_{c2} = -0.49417 \pm j0$

Form $D_3(p)$ the denominator of $T_{c3}(p)$

$D_3(p) = (p + 0.24709 - j0.96600)(p + 0.49417)(p + 0.24709 + j0.96600)$

$\qquad = (p + 0.49417)\left[(p + 0.24709)^2 + (0.96600)^2\right]$

$\qquad = (p + 0.49417)(p^2 + 0.49417p + 0.99421)$

$\qquad = p^3 + 0.98834p^2 + 1.2384p + 0.49131$

and the numerator of $T_{c3}(p)$ is 0.49131 so that $T_{c3}(0)$=1.

Form the denominator of $T_{c3}(p) T_{c3}(-p)$ on the $j\omega$ axis.

$D_3(p)D_3(-p) = (p^3 + 0.98834p^2 + 1.2384p + 0.49131)$

$\qquad\qquad\qquad\qquad \times (-p^3 + 0.98834p^2 - 1.2384p + 0.49131)$

$D_3(j\omega)D_3(-j\omega) = [-p^6 - 1.5p^4 - 0.562p^2 + 0.241]_{p=j\omega}$

$\qquad\qquad\qquad = \omega^6 - 1.5\omega^4 + 0.562\omega^2 + 0.241$

Compare $T_{c3}(p)T_{c3}(-p)$ to the numerator and denominator formed from one plus ε^2 times the Chebyshev polynomial.

$1 + \varepsilon^2 C_3^2(\omega) = 1 + \varepsilon^2(4\omega^3 - 3\omega)^2 = 16\varepsilon^2\left(\omega^6 - 1.5\omega^4 + \frac{9}{16}\omega^2 + \frac{1}{16\varepsilon^2}\right)$

$\qquad\qquad = (1/0.2414)(\omega^6 - 1.5\omega^4 + 0.5625\omega^2 + 0.2414)$

Problem 303 Ref Table 307. Derive $D_2(p)$ and $D_2(-p)$ for n=2, 1dB ripple
Problem 304 Derive $D_4(p)$ and $D_4(-p)$ for n=4, 1dB ripple.

Chebyshev Polynomials of the first Kind

The definition of $C_n(\omega)$ is $C_n(\omega) = \cos(n \cos^{-1}\omega)$. The recursion equation allows for calculation of the $C_n(\omega)$. The recursion relation for the $C_n(\omega)$ is $C_n(\omega) = 2\omega C_{n-1}(\omega) - C_{n-2}(\omega)$.

However calculation via trigonometry is more revealing.
The cos of the angle θ whose cos is ω is ω.

The sin of the angle θ whose cos is ω is $\sqrt{1-\omega^2}$

check : $\cos\theta = \omega$ *and* $\sin\theta = \sqrt{1-\omega^2}$ $\rightarrow$ $\cos^2\theta + \sin^2\theta = 1$

$C_0(\omega) = \cos(0\,arc\,\cos\,\omega) = \cos 0 = 1$

$C_1(\omega) = \cos(1\,arc\,\cos\,\omega) = \omega$

$C_2(\omega) = \cos(2\,arc\,\cos\,\omega) = \cos(arc\,\cos\,\omega + arc\,\cos\,\omega)$

$\qquad = \cos(arc\,\cos\,\omega)\cos(arc\,\cos\,\omega) - \sin(arc\,\cos\,\omega)\sin(arc\,\cos\,\omega)$

$\qquad = \omega \times \omega - \sqrt{1-\omega^2} \times \sqrt{1-\omega^2} = \omega^2 - (1-\omega^2) = 2\omega^2 - 1$

$\qquad$ *and so forth*

n	Table 306 Chebyshev polynomials $C_n(\omega)$
0	1
1	ω
2	$2\omega^2 - 1$
3	$4\omega^3 - 3\omega$
4	$8\omega^4 - 8\omega^2 + 1$
5	$16\omega^5 - 20\omega^3 + 5\omega$
6	$32\omega^6 - 48\omega^4 + 18\omega^2 - 1$
7	$64\omega^7 - 112\omega^5 + 56\omega^3 - 7\omega$
8	$128\omega^8 - 256\omega^6 + 160\omega^4 - 32\omega^2 + 1$
9	$256\omega^9 - 576\omega^7 + 432\omega^5 - 120\omega^3 + 9\omega$
10	$512\omega^{10} - 1280\omega^8 + 1120\omega^6 - 400\omega^4 + 50\omega^2 - 1$
11	$1024\omega^{11} - 2816\omega^9 + 2816\omega^7 - 1232\omega^5 + 220\omega^3 - 11\omega$

Transfer function products for Chebyshev Polynomials of the first Kind

$$T_{cn}(p)T_{cn}(-p) = \frac{1}{D_n(p)D_n(-p)} = \frac{1}{1+\varepsilon^2 C_n^{\,2}(-jp)} = \frac{1}{1+\varepsilon^2 C_n^{\,2}(\omega)}$$

$$1+\varepsilon^2 C_0^{\,2}(\omega) = 1+\varepsilon^2(1)^2 = \varepsilon^2\left(1+\tfrac{1}{\varepsilon^2}\right)$$

$$1+\varepsilon^2 C_1^{\,2}(\omega) = 1+\varepsilon^2(\omega)^2 = \varepsilon^2\left(\omega^2 + \tfrac{1}{\varepsilon^2}\right)$$

$$1+\varepsilon^2 C_2^{\,2}(\omega) = 1+\varepsilon^2(2\omega^2-1)^2 = 4\varepsilon^2\left(\omega^4 - \omega^2 + \tfrac{1}{4\varepsilon^2}\right)$$

$$1+\varepsilon^2 C_3^{\,2}(\omega) = 1+\varepsilon^2(4\omega^3-3\omega)^2 = 16\varepsilon^2\left(\omega^6 - 1.5\omega^4 + \tfrac{9}{16}\omega^2 + \tfrac{1}{16\varepsilon^2}\right)$$

And so forth. The problem here is that $1+\varepsilon^2 C_n(\omega)$ is *not* factored into $D_n(p)D_n(-p)$.

n	Table 307 Chebyshev $D_n(p)$ 1.0 dB $\varepsilon = 0.50885$
1	$p+1.9652$
2	$p^2 +1.0977p +1.1025$
3	$p^3 +0.9883p^2 +1.2384p +0.4913$
4	$p^4 +0.9528p^3 +1.4539p^2 +0.7426p +0.2756$
5	$p^5 +0.9368p^4 +1.6888p^3 +0.9744p^2 +0.5805p +0.1228$
6	$p^6 +0.9283p^5 +1.9308p^4 +1.2021p^3 +0.9393p^2 +0.3071p +0.0689$

A table of transfer functions is required for each value of the ripple factor ε. Calculated for what values of ε? Instead of tables, given ε we prefer to calculate $D(p)$ as was done on page 55.

An estimate for n is as follows.

$$\text{(31)} \quad n \geq \frac{\cosh^{-1}\left[\dfrac{10^{0.1\beta\, dB}-1}{10^{0.1\alpha\, dB}-1}\right]^{\frac{1}{2}}}{\cosh^{-1}(\omega_2/\omega_1)}$$

Figure 301 LPF Specification

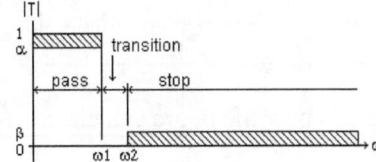

3.5 Inverse Chebyshev Approximation to a Low Pass Filter

Chebyshev used a recursive rule to create a second series of polynomials $U_n(\omega)$ whose value was constant when $\omega \leq 1$ (the low pass filter pass band), and whose value increased rapidly when $\omega > 1$ and fluctuated as $\pm \varepsilon/\sqrt{(1 \pm \varepsilon^2)}$ at the bottom of the attenuation band). The $U_n(\omega)$ are the *Chebyshev functions of the second kind*. The pass band ripples move to the attenuation band. However the $C_n(\omega)$ can be used when $1/\omega$ replaces ω.

$$(32) \quad M_{un}(\omega) = \sqrt{\cfrac{1}{1 + \cfrac{1}{\varepsilon^2 C_n^{\,2}\left(\frac{1}{\omega}\right)}}} = \sqrt{\frac{\varepsilon^2 C_n^{\,2}\left(\frac{1}{\omega}\right)}{1 + \varepsilon^2 C_n^{\,2}\left(\frac{1}{\omega}\right)}}$$

Transfer function

$$(33) \quad |z|^2 = z\bar{z}$$

if $\left|T_{un}(j\omega)\right| = M_{un}(\omega)$ *then*

$$(34) \quad T_{un}(j\omega)T_{un}(-j\omega) = M_{un}^{\,2}(\omega) = \frac{\varepsilon^2 C_n^{\,2}\left(\frac{1}{\omega}\right)}{1 + \varepsilon^2 C_n^{\,2}\left(\frac{1}{\omega}\right)}$$

To find T we recognize that

when $p = j\omega,\quad \omega = -jp = \dfrac{p}{j}$ *so that*

$$(35) \quad T_{un}(p)T_{un}(-p) = \frac{\varepsilon^2 C_n^{\,2}\left(\frac{j}{p}\right)}{1 + \varepsilon^2 C_n^{\,2}\left(\frac{j}{p}\right)}$$

The poles of $T_{un}(p)\,T_{un}(-p)$ are the zeros of the denominator $1 + \varepsilon^2 C^2$.

$$(36a) \quad 1 + \varepsilon^2 C_n^{\,2}(j/p) = 0 \;\rightarrow\; \varepsilon^2 C_n^{\,2}(j/p) = -1 = j^2 \;\rightarrow\; \varepsilon C_n(j/p) = j$$

$$(36b) \quad C_n\!\left(\frac{j}{p}\right) = 0 \pm j\frac{1}{\varepsilon} \quad \omega < 1$$

$$(37a) \quad from\ 28c \quad u = \frac{\pi}{2n}(2k-1) \quad and\ from\ 28d \quad v = \frac{1}{n}\sinh^{-1}\frac{1}{\varepsilon} \quad k = 1, 2, \ldots . n$$

$$(37b) \quad \frac{1}{p_k} = \frac{1}{\sigma_k + j\omega_k} = -\sinh v\,\sin\!\left(\frac{\pi}{2n}(2k-1)\right) + j\cosh v\,\cos\!\left(\frac{\pi}{2n}(2k-1)\right)$$

Here is how the p_k are calculated.

$$(38) \quad p_k = \sigma_k + j\omega_k = \frac{1}{u+jv} = \frac{1}{u+jv}\frac{u-jv}{u-jv} = \frac{u-jv}{u^2+v^2} = \frac{u}{u^2+v^2} - j\frac{v}{u^2+v^2}$$

The Chebyshev polynomials of the second kind are defined by the following equations, which we do not use .

(39) $\displaystyle U_n(x) = \frac{\sin((n+1)\theta)}{\sin\theta}$ $\qquad x = \cos\theta$

(40a) $U_0(\omega) = 1$

(40b) $U_1(\omega) = 2\omega$

(40c) $U_{n+1}(\omega) = 2\omega U_n(\omega) - U_{n-1}(\omega)$ $\qquad$ *recursion formula*

Or use trigonometry to calculate $U_n(x)$.

(41a) $\displaystyle U_0(x) = \frac{\sin((0+1)\theta)}{\sin\theta} = 1$

(41b) $\displaystyle U_1(x) = \frac{\sin((1+1)\theta)}{\sin\theta} = 2\cos\theta = 2x$

(41c) $\displaystyle U_2(x) = \frac{\sin((2+1)\theta)}{\sin\theta} = 3 - 4\sin^2\theta = 3 - 4(1-x^2) = 4x^2 - 1$

and so forth

n	Table 308 Chebyshev polynomials $U_n(\omega)$
0	1
1	$2x$
2	$4x^2 - 1$
3	$8x^3 - 4x$
4	$16x^4 - 12x^2 + 1$
5	$32x^5 - 32x^3 + 6x$
6	$64x^6 - 80x^4 + 24x^2 - 1$
7	$128x^7 - 192x^5 + 80x^3 - 8x$
8	$256x^8 - 448x^6 + 240x^4 - 40x^2 + 1$
9	$512x^9 - 1024x^7 + 672x^5 - 160x^3 + 10x$

4 Modern Filter Synthesis

4.1 Transfer Impedance Z Synthesis

Equations from Chapter 1 page 11 (Figure 102).

(16a) $V_S = (z_{11} + z_S)I_1 + z_{12}I_2$

(16b) $0 = z_{21}I_1 + (z_{22} + z_L)I_2$

response currents :

(17a) $\dfrac{I_1}{V_S} = \dfrac{(z_{22} + z_L)}{(z_{11} + z_S)(z_{22} + z_L) - z_{12}z_{21}} = \dfrac{z_{22} + z_L}{\Delta_{ZT}}$

(17b) $\dfrac{I_2}{V_S} = \dfrac{-z_{21}}{(z_{11} + z_S)(z_{22} + z_L) - z_{12}z_{21}} = -\dfrac{z_{21}}{\Delta_{ZT}}$

Figure 102 Terminated two-port

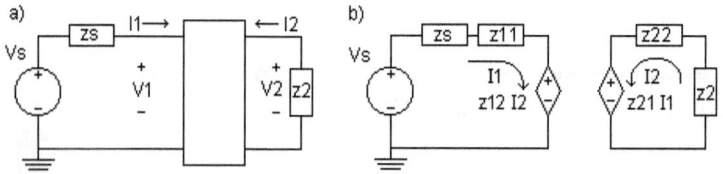

form 17b/17a

(1a) $\dfrac{I_2}{I_1} = -\dfrac{z_{21}}{z_{22} + z_2} \qquad z_2 = z_L$

(1b) $T(p) = \dfrac{V_2}{I_1} = -\dfrac{I_2 z_2}{I_1} = \dfrac{z_{21}z_2}{z_{22} + z_2}$

(2) $T_n(p) = \dfrac{1}{O_n(p) + E_n(p)}$

(3) $T_n(p) = \dfrac{\dfrac{1}{O_n(p)}}{1 + \dfrac{E_n(p)}{O_n(p)}} = \dfrac{z_{21}}{1 + z_{22}} \qquad (z_2 = 1)$

Hurwitz Polynomials

A polynomial H(p) is said to be Hurwitz if H(p) is real when p is real

The roots of H(p) have real parts, which are zero or negative.

Properties
1. The coefficients a_j of terms are all real and positive.
$$H(p) = a_n p^n + a_{n-1} p^{n-1} + ... + a_1 p^1 + a_0$$

2. The three types of roots H(p) can have are negative real, imaginary pairs, and complex pairs.
$$p = -\sigma_k, \quad p = \pm j\omega_k, \quad p = -\sigma_k \pm j\omega_k \quad (\sigma_k \text{ and } \omega_k > 0 \text{ and real})$$

$$H(p) = (p + \sigma_k)(p + j\omega_k)(p - j\omega_k)(p + \sigma_k + j\omega_k)(p + \sigma_k - j\omega_k)....$$
$$= (p + \sigma_k)(p^2 + \omega_k^2)(\{p + \sigma_k\}^2 + \omega_k^2)....$$
Clearly the coefficients of terms are positive.

3. The even and odd parts of H(p) have roots only on the $j\omega$ axis .

$$H(p) = O(p) + E(p)$$
example
$$H(p) = p^3 + 2p^2 + 2p + 1 = (p^3 + 2p) + (2p^2 + 1)$$
$$O(p) = p^3 + 2p = p(p^2 + 2) = p(p + j\sqrt{2})(p - j\sqrt{2})$$
$$E(p) = 2p^2 + 1 = 2(p + j\sqrt{1/2})(p - j\sqrt{1/2})$$

4. If H(p) is even or odd, then all of its roots are on the $j\omega$ axis.

5. The continued fraction expansion of the ratio O(p)/E(p) or E(p)/O(p) has all positive quotient terms. For example
$$\frac{O(p)}{E(p)} = \frac{p^3 + 2p}{2p^2 + 1} = \frac{p}{2} + \cfrac{1}{\cfrac{4}{3}p + \cfrac{1}{\cfrac{3}{2}p}}$$

Z Synthesis Examples

Butterworth n=5 Table 301 page 49, Spice 4021, 4022 pages 64, 65 The Butterworth filter denominator is Hurwitz, the numerator is even so divide by the odd polynomial O(p). Synthesize z_{22} (see Table 401 p68).

(4) $T_{bu5}(p) = \dfrac{1}{p^5 + 3.236p^4 + 5.236p^3 + 5.236p^2 + 3.236p + 1}$

$O_5(p) = p^5 + 5.236p^3 + 3.236p \qquad E_5(p) = 3.236p^4 + 5,236p^2 + 1$

(5) $T_{bu5}(p) = \dfrac{\dfrac{1}{O_5(p)}}{1 + \dfrac{E_5(p)}{O_5(p)}} = \dfrac{\dfrac{1}{p^5 + 5.236p^3 + 3.236p}}{1 + \dfrac{3.236p^4 + 5.236p^2 + 1}{p^5 + 5.236p^3 + 3.236p}} = \dfrac{z_{21}}{1 + z_{22}} \qquad (z_2 = 1)$

(6) $\dfrac{1}{z_{22}} = \dfrac{p^5 + 5.236p^3 + 3.236p}{3.236p^4 + 5.236p^2 + 1}$ *has components for* $R = 1, f_{-3dB} = \dfrac{1}{2\pi}$

Continued Fraction Expansion of equation 6.

$$\begin{array}{r} 0.309p \\ 3.236p^4 + 5.236p^2 + 1 \overline{\smash{\big)}\ p^5 + 5.236p^3 + 3.236p} \\ \underline{p^5 + 1.618p^3 + 0.309p} \\ 3.618p^3 + 2.927p \end{array}$$

Figure 403 Z_{21} Synthesis n=5

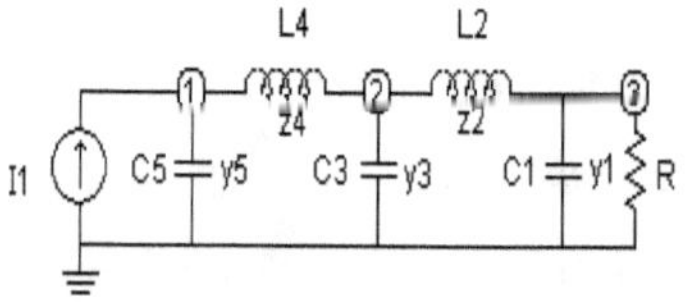

$$\begin{array}{r} 0.8944p \\ 3.618p^3 + 2.927p \overline{\smash{\big)}\ 3.236p^4 + 5.236p^2 + 1} \\ \underline{3.236p^4 + 2.618p^2 + 0} \\ 2.618p^2 + 1 \end{array}$$

$$\begin{array}{r} 1.382p \\ 2.618p^2 + 1 \overline{\smash{\big)}\ 3.618p^3 + 2.927p} \\ \underline{3.618p^3 + 1.382p} \\ 1.545p \end{array} \rightarrow \begin{array}{r} 1.694p \\ 1.545p \overline{\smash{\big)}\ 2.618p^2 + 1} \\ \underline{2.618p^2 + 0} \\ 1 \end{array} \rightarrow \begin{array}{r} 1.546p \\ 1 \overline{\smash{\big)}\ 1.546p} \\ \underline{1.546p} \\ 0p \end{array}$$

$$\frac{1}{z_{22}} = 0.309p + \cfrac{1}{0.894p + \cfrac{1}{1.382p + \cfrac{1}{1.694p + \cfrac{1}{1.546p}}}} = pC_1 + \cfrac{1}{pL_2 + \cfrac{1}{pC_3 + \cfrac{1}{pL_4 + \cfrac{1}{pC_5}}}}$$

Butterworth n=3 Table 301 page 49
Spice 4021, 4022 pages 64, 65

Figure 403 Z_{21} Synthesis n=3

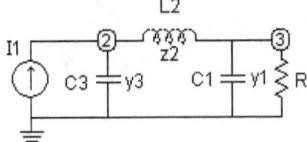

(7) $\dfrac{1}{z_{22Bu3}} = \dfrac{O_{Bu3}(p)}{E_{Bu3}(p)} = \dfrac{p^3 + 2p}{2p^2 + 1}$

$2p^2 + 1 \overline{\smash{)}\, p^3 + 2p} \;\rightarrow\; \tfrac{1}{2}p$

$\qquad \underline{p^3 + \tfrac{1}{2}p}$

$\qquad\qquad \tfrac{3}{2}p \overline{\smash{)}\, 2p^2 + 1} \;\rightarrow\; \tfrac{4}{3}p$

$\qquad\qquad\quad \underline{2p^2}$

$\qquad\qquad\qquad\qquad 1 \overline{\smash{)}\, \tfrac{3}{2}p} \;\rightarrow\; \tfrac{3}{2}p$

See Table 401 p68 $C_1 = \tfrac{1}{2}p = 0.5000 \quad L_2 = \tfrac{4}{3}p = 1.3333 \quad C_3 = \tfrac{3}{2}p = 1.5000$

Chebyshev $D_3(p)$ page 57, Spice 4031 page 66. Compare results to Table 402 n=3 page 68.

(8) $\dfrac{1}{z_{22Ch3}} = \dfrac{O_{Ch3}(p)}{E_{Ch3}(p)} = \dfrac{p^3 + 1.2384p}{0.9883p^2 + 0.4913}$

$0.9883p^2 + 0.4913 \overline{\smash{)}\, p^3 + 1.2384p} \;\rightarrow\; \tfrac{1}{0.9883}p$

$\qquad \underline{p^3 + 0.4971p}$

$\qquad\qquad 0.7413p \overline{\smash{)}\, 0.9883p^2 + 0.4913} \;\rightarrow\; \tfrac{0.9883}{0.7413}p$

$\qquad\qquad\quad \underline{0.9983p^2}$

$\qquad\qquad\qquad\qquad 0.4913 \overline{\smash{)}\, 0.7413p} \;\rightarrow\; \tfrac{0.7413}{0.4913}p$

$C_1 = \tfrac{1}{0.9883}p = 1.0118 \qquad L_2 = \tfrac{0.9883}{0.7413}p = 1.3332 \qquad C_3 = \tfrac{0.7413}{0.4913}p = 1.5089$

Bessel n=3 Table 304 page 52, Spice 4032 page 67. Compare results to Table 403 n=3 page 68.

(9) $\dfrac{1}{z_{22Be3}} = \dfrac{O_{Be3}(p)}{E_{Be3}(p)} = \dfrac{p^3 + 15p}{6p^2 + 15}$

$6p^2 + 15 \overline{\smash{)}\, p^3 + 15p} \;\rightarrow\; \tfrac{1}{6}p$

$\qquad \underline{p^3 + \tfrac{5}{2}p}$

$\qquad\qquad \tfrac{25}{2}p \overline{\smash{)}\, 6p^2 + 15} \;\rightarrow\; \tfrac{12}{25}p$

$\qquad\qquad\quad \underline{6p^2}$

$\qquad\qquad\qquad\qquad 15 \overline{\smash{)}\, \tfrac{25}{2}p} \;\rightarrow\; \tfrac{5}{6}p$

compare to Table 403 n = 3

$C_1 = \tfrac{1}{6}p = 0.1667$

$L_2 = \tfrac{12}{25}p = 0.4800$

$C_3 = \tfrac{5}{6}p = 0.8333$

Filter Transfer Function Plots **Butterworth n=5 page 62 and Butterworth n=3 page 63.** Compare calculated values to normalized LC component values for single terminated filters (Figures 402, 403) in Table 401 n= 3, n=5 page 68. R_0=1 ohm and f_{-3dB}=1/2π Hertz.

Figure 402 z_{21} n=5   **Figure 403 z_{21} n=3**

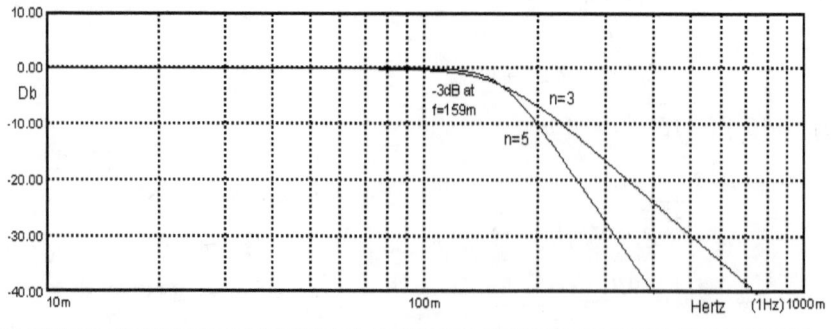

Spice program 4021 Z Synthesis

```
Fig4021.ckt  Butterworth Filters n=3,5
I1 1 0 AC 1 0     ; Figure 403
C3 1 0 1.5
L2 1 2 1.333
C1 2 0 0.5
R1 2 0 1


I11 11 0  AC 1 0
C15 11 0     1.546   ; Figure 402
L14 11 12    1.694
C13 12 0     1.382
L12 12 13    0.894
C11 13 0     0.309
R11 13 0     1
*-3dB at omega=1 or f=1/2pi
.AC DEC 201 0.01 1       ; 10mHz=0.01Hz to 1000mHZ=1Hz
.TEMP 27
.PLOT AC VDB(2) VDB(13) -40,10
.end
```

Figure 40211 Butterworth, n=3 and n=5, z_{21} transfer function

Butterworth n=5 page 62 and Butterworth n=3 page 63. Calculated values are scaled to 600 ohms and 100 KHz on page 79.

Compare calculated values to normalized LC component values for single terminated filters (Figures 402, 403) are taken from Table 401 n= 3, n=5 page 68.

Spice program 4022 Z Synthesis

Fig4022.ckt   Butterworth Filters n=3,5

```
I1 1 0 AC 0.00167  0      ; Figure 403
C3 1 0    3980p          ;1.5
L2 1 2    1273u          ;1.333
C1 2 0    1327p          ;0.5
R1 2 0    600            ;1

I11 11 0  AC 0.00167 0   ; Figure 402
C15 11 0   4100p         ;1.546
L14 11 12  1619u         ;1.694
C13 12 0   3667p         ;1.381
L12 12 13  854u          ;0.894
C11 13 0   820p          ;0.309
R11 13 0   600           ;1
```

```
* -3dB at f=100K hertz
*.PLOT AC VP(2) VP(3) -150,100
.AC DEC 201 10000 1e+006
.TEMP 27
.PLOT AC VDB(2) VDB(13) -40,10
.end
```

Figure 40221 Butterworth, n=3 and n=5, z_{21} transfer function

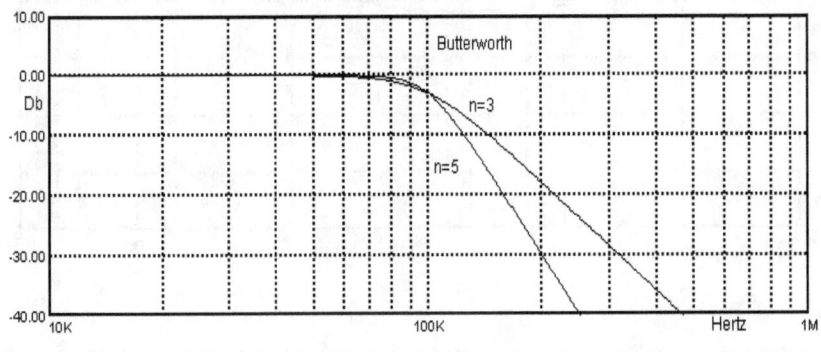

Chebyshev $D_3(p)$ page 63 Compare calculated values to normalized LC component values for single terminated filters (Figure 403) are taken from Table 402 n= 3 page 68. R_0=1 ohm and f_{-3dB}=1/2π Hertz.

Figure 403 z_{21} n=3

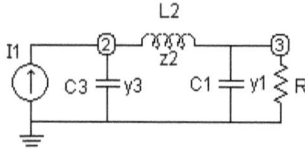

Spice program 4031 Z Synthesis

Fig4031.ckt   Chebyshev Filter n=3

```
I1 1 0 AC 1 0
C3 1 0 1.5089
L2 1 2 1.3332
C1 2 0 1.1018
R1 2 0 1

* -3db at f=1/2pi Hertz
.AC DEC 201 0.01 1
.TEMP 27
.PLOT AC VDB(2) -20,5
.PRINT AC VDB(2)
.end
```

Figure 40311 Chebyshev, n=3, z_{21} transfer function

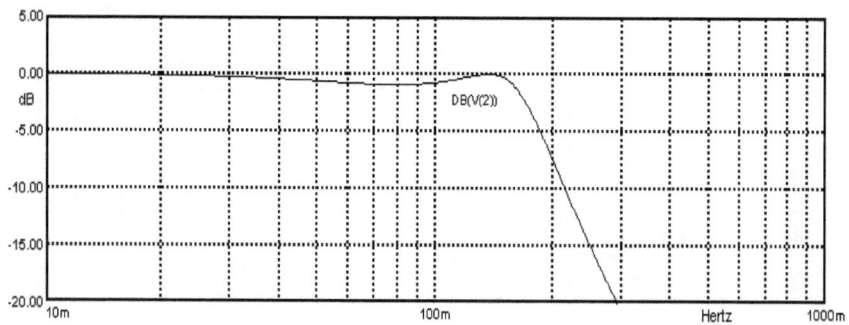

Bessel   n=3   page   63 Compare calculated values to normalized LC component values for single terminated filters (Figure 403) are taken from Table 403 n= 3, page 68. R_0=1 ohm and f_{-3dB}=1/2π Hertz.

Figure 403 z_{21} n=3

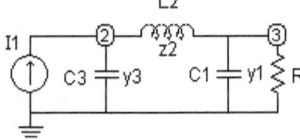

Spice program 4032 Z Synthesis

Fig4032.ckt   Bessel Filter n=3

```
I1 1 0 AC 1 0
C3 1 0    0.83333     ; Figure 403
L2 1 2    0.4800
C1 2 0    0.1667
R1 2 0    1

* -3db at f=1/2pi Hertz
.AC DEC 201 0.01 1
.TEMP 27
.PLOT AC VDB(2) -20,5
.PRINT AC VDB(2)
.end
```

Figure 40321 Bessel, n=3, z_{21} transfer function

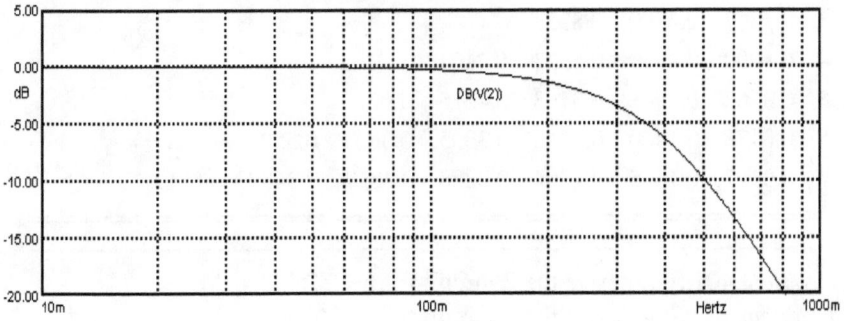

Table 401 Butterworth Low Pass Filters, Z Synthesis						
n	C_1	L_2	C_3	L_4	C_5	L_6	C_7
1	1.0000						
2	0.7071	1.4142					
3	0.5000	1.3333	1.5000				
4	0.3827	1.0824	1.5772	1.5307			
5	0.3090	0.8944	1.3820	1.6944	1.5451		
6	0.2588	0.7579	1.2016	1.5529	1.7593	1.5529	
7	0.2225	0.6560	1.0550	1.3972	1.6588	1.7988	1.5576

Table 402 Chebyshev Low Pass Filters, 1dB ripple, Z Synthesis						
n	C_1	L_2	C_3	L_4	C_5	L_6	C_7
1	0.5088						
2	0.9110	0.9957					
3	1.0118	1.3332	1.5088				
4	1.0495	1.4126	1.9093	1.2817			
5	1.0674	1.4441	1.9938	1.5908	1.6652		
6	1.0773	1.4601	2.0270	1.6507	2.0491	1.3457	
7	1.0832	1.4694	2.0437	1.6736	2.1192	1.6489	1.7118

Table 403 Bessel Low Pass Filters, Z synthesis						
n	C_1	L_2	C_3	L_4	C_5	L_6	C_7
1	1.0000						
2	0.3333	1.0000					
3	0.1667	0.4800	0.8333				
4	0.1000	0.2899	0.4627	0.7101			
5	0.0667	0.1948	0.3103	0.4215	0.6231		
6	0.0476	0.1400	0.2246	0.3005	0.3821	0.5595	
7	0.0357	0.1055	0.1704	0.2288	0.2827	0.3487	0.5111

Problem 401 Synthesize the Z circuit for n=2 Table 401
Problem 402 Synthesize the Z circuit for n=6 Table 401
Problem 403 Synthesize the Z circuit for n=4 Table 402
Problem 404 Synthesize the Z circuit for n=6 Table 402
Problem 405 Synthesize the Z circuit for n=5 Table 403

4.2 Darlington Insertion Loss Synthesis

One of Darlington's[8] ingenious inventions was converting the filter transfer function specification to synthesis of an input impedance, which turns out to be the straightforward procedure developed in upcoming paragraphs. Darlington adapted the reflection coefficient idea to achieve his method. Darlington made a continuous fraction expansion of the input impedance $z_{in}(p)$ to create double terminated filters with source and load r_0.

The maximum power the source can deliver and the power delivered to the load are as follows (Figure 102 page 60).

$$(10a) \quad P_{smax} = \frac{|V_S(j\omega)|^2}{4R_s} \qquad (10b) \quad P_{load} = \frac{|V_2(j\omega)|^2}{R_L}$$

The input impedance at port 1 is z_{in}.

$$(11) \quad z_{in} = R_{in} + jX_{in} \quad \rightarrow \quad I_1(p) = \frac{V_S(p)}{R_S + z_{in}} \quad \rightarrow \quad I_1(j\omega) = \frac{V_S(j\omega)}{R_S + z_{in}(j\omega)}$$

$$(12a) \quad \text{since the filter is lossless } P_{in} = P_{load} \quad \rightarrow \quad R_{in}|I_1(j\omega)|^2 = \frac{|V_2(j\omega)|^2}{R_L}$$

$$(12b) \quad R_{in}\frac{|V_S(j\omega)|^2}{|R_S + z_{in}(j\omega)|^2} = \frac{|V_2(j\omega)|^2}{R_L}$$

Define the voltage transmission ratio through the filter (equation 12b).

$$(13) \quad |T_n(j\omega)|^2 = \frac{|V_2(j\omega)|^2}{|V_S(j\omega)|^2} = \frac{R_{in}R_L}{|R_S + z_{in}(j\omega)|^2} = \frac{R_{in}R_L}{|R_S + R_{in} + jX_{in}|^2} = \frac{R_{in}R_L}{(R_S + R_{in})^2 + X_{in}^2}$$

Define the power transmission ratio $t_n(p)$ through the filter.

$$(14a) \quad |t_n(j\omega)|^2 = \frac{P_{load}}{P_{smax}} = \frac{4R_s}{R_L}\frac{|V_2(j\omega)|^2}{|V_S(j\omega)|^2} = \frac{4R_s}{R_L}|T_n(j\omega)|^2$$

$$(14b) \quad t_n(p)t_n(-p) = \frac{4R_S}{R_L}T_n(p)T_n(-p)$$

---

[8] S. Darlington, "Synthesis of Reactance 4-Poles which Produce Prescribed Insertion Loss Characteristics," J. Math. Phys., **18**, 1939, 257-353.

The reflection coefficient ρ is defined as the ratio of power returned to the source divided by the power delivered by the source. The LC filter is lossless.

$$(15a) \quad |\rho(j\omega)|^2 = \frac{P_{returned}}{P_{smax}} = \frac{P_{smax} - P_{load}}{P_{smax}} = 1 - \frac{P_{load}}{P_{smax}} = 1 - \frac{4R_S}{R_L}|T_n(j\omega)|^2$$

$$(15b) \quad |\rho(j\omega)|^2 = 1 - \frac{4R_S}{R_L}\frac{R_{in}R_L}{(R_S + R_{in})^2 + X_{in}^2} = 1 - \frac{4R_{in}R_S}{(R_S + R_{in})^2 + X_{in}^2}$$

$$(15c) \quad |\rho(j\omega)|^2 = \frac{(R_S + R_{in})^2 + X_{in}^2 - 4R_{in}R_S}{(R_S + R_{in})^2 + X_{in}^2} = \frac{(R_S - R_{in})^2 + X_{in}^2}{(R_S + R_{in})^2 + X_{in}^2} = \frac{|R_S - z_{in}|^2}{|R_S + z_{in}|^2}$$

$$(16a) \quad \rho(p) = \frac{R_S - z_{in}(p)}{R_S + z_{in}(p)} \quad \rightarrow \quad (16b) \quad \frac{z_{in}(p)}{R_S} = \frac{1 \pm \rho(p)}{1 \mp \rho(p)} = \frac{N_z(p)}{D_z(p)}$$

$Z_{in}(p)$ has two possible physical realizations ($N_z(p)/D_z(p)$ or $D_z(p)/N_z(p)$). Darlington proved that $z_{in}(p)$ can always be realized by lossless LC circuits terminated in a resistor R, and that the two circuits are duals.

Given a T(p), synthesis of a Darlington filter circuit first finds t(p) from T(p). Then the process finds ρ(p), and finally $z_{in}(p)$. Here is an example.

Example Darlington Butterworth Filter, n=3, Spice 4062 page 74.
(equation 5 page 47, Table 301 n=3 page 49, Figure 404).

$$(17) \quad T_n(j\omega)T_n(-j\omega) = \frac{1}{1+\omega^{2n}}$$

$$(18) \quad t_n(j\omega)t_n(-j\omega) = \frac{R_S}{R_L}T_n(j\omega)T_n(-j\omega) = \frac{1}{1+\omega^{2n}}$$

$$(19) \quad \rho_n(j\omega)\rho_n(-j\omega) = 1 - t_n(j\omega)t_n(-j\omega) \quad \textit{from 15a}$$

$$= 1 - \frac{1}{1+\omega^{2n}} = \frac{\omega^{2n}}{1+\omega^{2n}} = \omega^n T_n(j\omega) \times \omega^n T_n(-j\omega)$$

$$(20a) \quad \rho_n(p) = p^n T_n(p)$$

$$(20b) \quad z_{in}(p) = R_0\frac{1+\rho(p)}{1-\rho(p)} = R_0\frac{1 + p^n T_n(p)}{1 - p^n T_n(p)}$$

When n=3 (compare to Table 404 n=3 page 76)

$$(21) \quad z_{in3} = 1 - \frac{1 + \dfrac{p^3}{p^3 + 2p^2 + 2p + 1}}{1 - \dfrac{p^3}{p^3 + 2p^2 + 2p + 1}} = \frac{2p^3 + 2p^2 + 2p + 1}{2p^2 + 2p + 1}$$

$$(22) \quad z_{in3} = p + \frac{1}{2p + \dfrac{1}{p + \dfrac{1}{1}}} = pL_1 + \frac{1}{pC_2 + \dfrac{1}{pL_3 + \dfrac{1}{R_0}}}$$

Figure 404   **Figure 405**

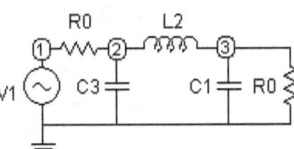

Butterworth   Define ρ using *admittance* (Figure 405).

$$(23a) \quad \rho_n(p) = p^n T_n(p) \qquad\qquad (23b) \quad \rho(p) = \frac{y - g_0}{y + g_0}$$

$$(23c) \quad y_n = g_0 \frac{1 + \rho_n(p)}{1 - \rho_n(p)} = g_0 \frac{1 + p^n T_n(p)}{1 - p^n T_n(p)}$$

$$(24a) \quad y_3 = 1 - \frac{1 + \dfrac{p^3}{p^3 + 2p^2 + 2p + 1}}{1 - \dfrac{p^3}{p^3 + 2p^2 + 2p + 1}} = \frac{2p^3 + 2p^2 + 2p + 1}{2p^2 + 2p + 1}$$

$$(24b) \quad y_3 = p + \frac{1}{2p + \dfrac{1}{p + \dfrac{1}{1}}} = pC_1 + \frac{1}{pL_2 + \dfrac{1}{pC_3 + \dfrac{1}{R_0}}}$$

Problem 406 Synthesize the Darlington circuit for n=4 Table 404
Problem 407 Synthesize the Darlington circuit for n=3 Table 405
Problem 408 Synthesize the Darlington circuit for n=5 Table 405
Problem 409 Synthesize the Darlington circuit for n=4 Table 406
Problem 410 Synthesize the Darlington circuit for n=7 Table 406

Analog Filter Design

Example Darlington Bessel n=2 Table 304 page 52. Table 406 page 76

(25) $T_{be2} = \dfrac{B_{e2}(0)}{B_{e2}(p)} = \dfrac{3}{p^2+3p+3}$

(26) $B_{e2}(p)B_{e2}(-p) = \left(p^2+3p+3\right)\left(p^2-3p+3\right) = p^4-3p^2+9$

(27) $t_2(p)t_2(-p) = T_{be2}(p)T_{be2}(-p) = \dfrac{9}{B_{e2}(p)B_{e2}(-p)} = \dfrac{9}{p^4-3p^2+9}$

(28a) $\rho_2(p)\rho_2(-p) = 1-t_2(p)t_2(-p) = 1-\dfrac{9}{B_{e2}(p)B_{e2}(-p)} = 1-\dfrac{9}{p^4-3p^2+9}$

$= \dfrac{p^4-3p^2+9-9}{p^4-3p^2+9} = \dfrac{p^4-3p^2}{p^4-3p^2+9} = \dfrac{p(p+\sqrt{3})p(p-\sqrt{3})}{p^4-3p^2+9}$

(28b) $\rho_2(p) = \dfrac{p(p+\sqrt{3})}{p^2+3p+3}$

(29) $\dfrac{z_{in}}{R_S} = \dfrac{1+\rho_2(p)}{1-\rho_2(p)} = \dfrac{(p^2+3p+3)+p(p+\sqrt{3})}{(p^2+3p+3)-p(p+\sqrt{3})} = \dfrac{2p^2+(3+\sqrt{3})p+3}{(3-\sqrt{3})p+3}$

Continued fraction expansion produces the circuit (compare to Table 406 n=2 page 76).

(30)

$$
\begin{array}{r}
1.268p+3\overline{\smash{\big)}\,2p^2+4.732p+3} \quad \rightarrow \quad \tfrac{2}{1.268}p = 1.5773p \\
\underline{2p^2+4.732p} \\
3\overline{\smash{\big)}\,1.268p+3} \quad \rightarrow \quad \tfrac{1.268}{3}p = 0.4226p \\
\underline{1.268p} \\
3\overline{\smash{\big)}\,3} \quad \rightarrow \quad 1
\end{array}
$$

Reference Spice program 4061, Figure 40611 page 75. Plot the results.
Problem 411 Write a Spice program for Darlington Bessel n=2.
Problem 412 Write a Spice program for Darlington Chebyshev n=3.

Example Darlington Chebyshev n=3 Table 307 page 57.

Note: Darlington circuits only for n odd.

$$(31) \quad T_{c3} = \frac{0.4913}{p^3 + 0.9883p^2 + 1.2384p + 0.4913} = \frac{0.4913}{D_{c3}(p)}$$

$$(32) \quad D_{c3}(p)D_{c3}(-p) = -p^6 - 1.5p^4 - \tfrac{9}{16}p^2 + (0.4913)^2 \quad \textit{from page 55}$$

$$(33) \quad t_3(p)t_3(-p) = T_{c3}(p)T_{c3}(-p) = \frac{(0.4913)^2}{D_{c3}(p)D_{c3}(-p)}$$

$$(34a) \quad \rho_2(p)\rho_2(-p) = 1 - t_2(p)t_2(-p) = 1 - \frac{(0.4913)^2}{-p^6 - 1.5p^4 - \tfrac{9}{16}p^2 + (0.4913)^2}$$

$$= \frac{-p^6 - 1.5p^4 - \tfrac{9}{16}p^2 + (0.4913)^2 - (0.4913)^2}{-p^6 - 1.5p^4 - \tfrac{9}{16}p^2 + (0.4913)^2}$$

$$= \frac{-p^2(p^4 + 1.5p^2 - \tfrac{9}{16})}{-p^6 - 1.5p^4 - \tfrac{9}{16}p^2 + (0.4913)^2}$$

$$(34b) \qquad \rho_2(p) = \frac{p(p^2 + \tfrac{3}{4})}{D_{c3}(p)}$$

$$(35) \quad \frac{z_{in}}{R_S} = \frac{1 + \rho_2(p)}{1 - \rho_2(p)} = \frac{D_{c3}(p) + p(p^2 + \tfrac{3}{4})}{D_{c3}(p) - p(p^2 + \tfrac{3}{4})}$$

$$= \frac{2p^3 + 0.9883p^2 + 1.9884p + 0.4913}{0.9883p^2 + 0.4884p + 0.4913}$$

Continued fraction expansion produces the circuit (compare to Table 405 n=3 page 76).

$$(36) \quad 0.9883p^2 + 0.4884p + 0.4913 \overline{\smash{\big)}\ 2p^3 + 0.9883p^2 + 1.9884p + 0.4913} \quad \rightarrow \quad \tfrac{2}{0.9883}p$$

$$\frac{2p^2 + 0.9883p^2 + 0.9942p}{0.9942p + 0.4913} \qquad = 2.0237p$$

$$0.9942p + 0.4913 \overline{\smash{\big)}\ 0.9883p^2 + 0.4884p + 0.4913} \qquad \rightarrow \quad \tfrac{0.9883}{0.9942}p = 0.9941p$$

$$\frac{0.9883p^2 + 0.4884p}{0.4913 \overline{\smash{\big)}\ 0.9942p + 0.4913}} \quad \rightarrow \quad \tfrac{0.9942}{0.4913}p = 2.0237p$$

$$0.4913 \overline{\smash{\big)}\ 0.4913} \rightarrow 1$$

Compare calculated values to normalized LC component values for double terminated filters (Figures 406, 405) taken from Table 404 n= 3, n=5 page 76. R_0=1 ohm and f_{-3dB}=1/2π Hertz.

Figure 406 z_{21} n=5   **Figure 405 z_{21} n=3**

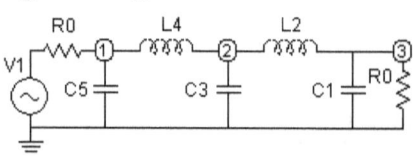 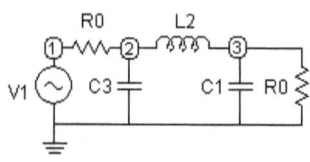

Spice program 4062 Darlington Synthesis

```
Fig4062.ckt    Filters n=3,5
V1 1 0  AC 1  0   ; volts
R1 1 2 1
C3 2 0 1
L2 2 3 2
C1 3 0 1
R2 3 0 1


R11 1  12 1
C15 12  0 0.618
L14 12 13 1.618
C13 13  0 2
L12 13 14 1.618
C11 14  0 0.618
R12 14  0 1


.AC DEC 201 0.01 1        ; 10mHz=0.01Hz to 1000mHZ=1Hz
.TEMP 27
.PLOT AC VDB(3) VDB(14) -50,0
.end
```

Figure 40621 Butterworth, n=3 and n=5, Darlington Synthesis

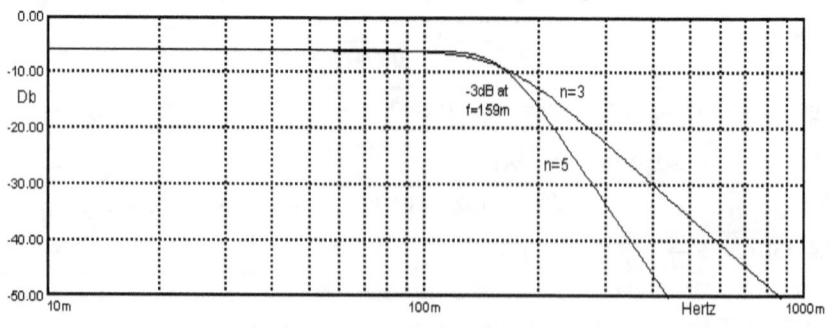

Compare calculated values to normalized LC component values for double terminated filters (Figure 406) taken from Tables 404, 405, 406 n=5 page 76. $R_0=1\Omega$. $f_{-3dB}=1/2\pi$ Hertz.

Figure 40611 Chebyshev, Butterworth, Bessel n=5, magnitude

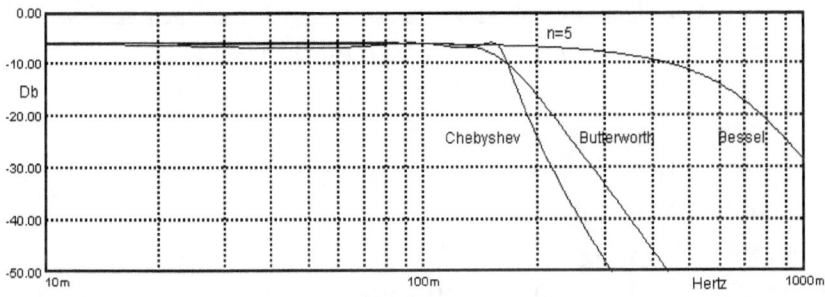

Spice program 4061 Darlington Synthesis

```
Fig4061.ckt    Filters n=5
V11 11 0  AC 1  0     ; volts
R10 11 12 1    ;Butterworth
C11 12  0 0.618
L12 12 13 1.618
C13 13  0 2
L14 13 14 1.618
C15 14  0 0.618
R16 14  0 1

V21 21 0  AC 1  0
R20 21 22 1    ;Chebyshev
C21 22  0 2.135
L22 22 23 1.091
C23 23  0 3.001
L24 23 24 1.091
C25 24  0 2.135
R26 24  0 1
V31 31 0  AC 1  0
R30 31 32 1    ;Bessel
C31 32  0 0.072
L32 32 33 0.209
C33 33  0 0.331
L34 33 34 0.458
C35 34  0 0.930
R36 34  0 1
.AC DEC 201 0.01 1
.TEMP 27
.PLOT AC VDB(14) VDB(24) VDB(34) -50,0
.end
```

Repetition of the examples on pages 70 to 73 for various n produced the normalized LC component values for double terminated filters (such as those in Figures 404, 405, 406) in Tables 404, 405, 406. R_0=1 ohm and f_{-3dB}=1/2π Hertz.

Table 404 *Butterworth Low Pass Filters, Darlington Synthesis*

n	C_1	L_2	C_3	L_4	C_5	L_6	C_7
1	2.0000						
2	1.4142	1.4142					
3	1.0000	2.0000	1.0000				
4	0.7654	1.8478	1.8478	0.7654			
5	0.6180	1.6180	2.0000	1.6180	0.6180		
6	0.5176	1.4142	1.9319	1.9319	1.4142	0.5176	
7	0.4450	1.2470	1.8019	2.0000	1.8019	1.2470	0.4450

Table 405 *Chebyshev Low Pass Filters, 1dB ripple, Darlington Synthesis*

n	C_1	L_2	C_3	L_4	C_5	L_6	C_7	L_8	C_9
1	1.0177								
3	2.0236	0.9941	2.0236						
5	2.1349	1.0911	3.0009	1.0911	2.1349				
7	2.1666	1.1115	3.0936	1.1735	3.0936	1.1115	2.1666		
9	2.1797	1.1192	3.1214	1.1897	3.1746	1.1897	3.1414	1.1192	2.1797

Table 406 *Bessel Low Pass Filters, Darlington Synthesis*

n	C_1	L_2	C_3	L_4	C_5	L_6	C_7
1	2.000						
2	1.577	0.423					
3	1.255	0.553	0.192				
4	1.060	0.512	0.318	0.110			
5	0.930	0.458	0.331	0.209	0.072		
6	0.838	0.412	0.316	0.236	0.148	0.051	
7	0.768	0.374	0.294	0.238	0.178	0.110	0.038

4.3 High Pass, Band Pass, Band Reject Filters

Low pass filter designs are converted by *frequency transformation* to any other form of filter. Frequency transformation changes the components of a low pass filter to the components of high pass, band pass, and band reject filters. Frequency transformations scale frequency. In addition the transformed circuits need to be scaled for impedance level.

Figure 407 Frequency Transformations

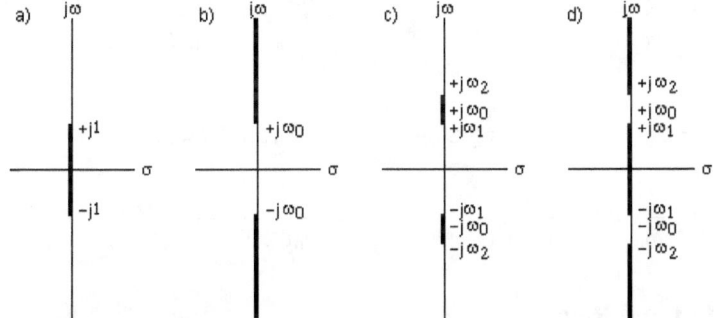

Low Pass to High Pass

$$(18) \quad p_n = \frac{\omega_0}{p}$$

$$(19a) \quad L_n p_n = L_n \frac{\omega_0}{p} \equiv \frac{1}{pC_h} \quad \Rightarrow \quad C_h = \frac{1}{\omega_0 L_n}$$

$$(19b) \quad C_n p_n = C_n \frac{\omega_0}{p} \equiv \frac{1}{pL_h} \quad \Rightarrow \quad L_h = \frac{1}{\omega_0 C_n}$$

Figure 408

Low Pass	High Pass	Band Pass		Band Reject	
L_n	C_h	L_s	C_s	L_p C_p	
C_n	L_h	L_p C_p		L_s	C_s

Low Pass to Band Pass

(20) $\quad p_n = Q\left(\dfrac{p}{\omega_0} + \dfrac{\omega_0}{p}\right)$

where $Q = \dfrac{\omega_0}{bandwidth} = \dfrac{\omega_0}{\omega_2 - \omega_1}$

(21a) $\quad L_n P_n = L_n Q\left(\dfrac{p}{\omega_0} + \dfrac{\omega_0}{p}\right) \equiv pL_S + \dfrac{1}{pC_S}$ *(series $L_S C_S$)*

$\quad L_S = L_n \dfrac{Q}{\omega_0} = L_n \dfrac{1}{\omega_2 - \omega_1} \qquad C_S = \dfrac{1}{\omega_0 L_n Q} = \dfrac{\omega_2 - \omega_1}{\omega_0^2 L_n}$

(21b) $\quad C_n P_n = C_n Q\left(\dfrac{p}{\omega_0} + \dfrac{\omega_0}{p}\right) \equiv pC_P + \dfrac{1}{pL_P}$ *(parallel $L_P C_P$)*

$\quad C_P = C_n \dfrac{Q}{\omega_0} = C_n \dfrac{1}{\omega_2 - \omega_1} \qquad L_P = \dfrac{1}{\omega_0 C_n Q} = \dfrac{\omega_2 - \omega_1}{\omega_0^2 C_n}$

Low Pass to Band Reject

(22) $\quad p_n = \dfrac{1}{Q\left(\dfrac{p}{\omega_0} + \dfrac{\omega_0}{p}\right)}$

where $Q = \dfrac{\omega_0}{bandwidth} = \dfrac{\omega_0}{\omega_2 - \omega_1}$

(23a) $\quad L_n P_n = L_n \dfrac{1}{Q\left(\dfrac{p}{\omega_0} + \dfrac{\omega_0}{p}\right)} \equiv \dfrac{1}{pC_p + \dfrac{1}{pL_p}}$ *(parallel LC)*

$\quad C_p = \dfrac{Q}{\omega_0 L_n} = \dfrac{1}{L_n(\omega_2 - \omega_1)} \qquad L_p = \dfrac{L_n}{\omega_0 Q} = L_n \dfrac{\omega_2 - \omega_1}{\omega_0^2}$

(23b) $\quad C_n P_n = C_n \dfrac{1}{Q\left(\dfrac{p}{\omega_0} + \dfrac{\omega_0}{p}\right)} \equiv \dfrac{1}{pL_s + \dfrac{1}{pC_s}}$ *(series LC)*

$\quad L_s = \dfrac{Q}{\omega_0 C_n} = \dfrac{1}{C_n(\omega_2 - \omega_1)} \qquad C_s = C_n \dfrac{1}{\omega_0 Q} = C_n \dfrac{\omega_2 - \omega_1}{\omega_0^2}$

An example transforming the circuit impedance and the -3dB frequency

$$\omega_{normalized} = \omega_n = \frac{\omega}{\omega_0} \quad \rightarrow \quad \omega_0 \;\; is \;\; the \;\; design \;\; -3dB \;\; frequency$$

$$z_{normalized} = z_n = \frac{z}{R_0} \quad \rightarrow \quad R_0 \;\; is \;\; the \;\; design \;\; impedance \;\; level$$

Scale frequency.

$$\omega_n L_n = \omega_0 L \qquad \qquad \omega_n C_n = \omega_0 C$$

$$L = \frac{\omega_n}{\omega_0} L_n \qquad \qquad C = \frac{\omega_n}{\omega_0} C_n$$

Scale impedance $Z = R_0 Z_n$

$$R = R_0 R_n \qquad L = \frac{\omega_n}{\omega_0} L_n R_0 \qquad C = \frac{\omega_n}{\omega_0} \frac{C_n}{R_0}$$

when $R_n = 1,$ and $\omega_n = 1 = 2\pi f_n = 2\pi(1/2\pi)$

$$R = R_0 \qquad L = \frac{L_n R_0}{\omega_0} \qquad C = \frac{C_n}{\omega_0 R_0}$$

Transform the Butterworth n=5 low pass filter circuit impedance from 1 ohm to 600 ohms and the -3dB frequency from $1/2\pi$ Hertz to 100,000 Hertz.

(a) $R = 600\Omega$

(d) $C_3 = \dfrac{1.382}{0.309} C_1 = 3667\,pF$

(b) $C_1 = 0.309 \cdot \dfrac{2\pi(1/2\pi)}{2\pi(10^5)600} = 820\,pF$

(e) $L_4 = \dfrac{1.694}{0.894} L_2 = 1619\mu H$

(c) $L_2 = 0.894 \cdot \dfrac{600}{2\pi 10^5} = 854\mu H$

(f) $C_5 = \dfrac{1.546}{0.309} C_1 = 4100\,pF$

The filter transfer function is plotted by Spice program 4022 on page 65.

Problem 413 Ref Table 404 n=5. Convert to high pass, 100KHz, 600Ω.
Problem 414 Ref Table 404 n=3. Convert to high pass, 1MHz, 50Ω.

Design – Z Synthesis

1 Write a specification

2 Select type of approximation - Butterworth, Bessel, Chebyshev, Inverse Chebyshev.

3 Calculate n

4 Form $T_n(p)$ using Table 301 page 49, or Table 304 page 52, or Table 307 page 57. Add a numerator q so that $T_n(0)=1$.

5 Form $T_n(p) = \dfrac{N_n(p)}{D_n(p)} = \dfrac{q}{O(p)+E(p)} = \dfrac{q\dfrac{1}{O(p)}}{1+\dfrac{E(p)}{O(p)}}$

6 Perform a continued fraction expansion on $\dfrac{E(p)}{O(p)}$ or $\dfrac{O(p)}{E(p)}$ to produce components.

7 Draw a schematic.

Design – Darlington Synthesis

1 Write a specification

2 Select type of approximation - Butterworth, Bessel, Chebyshev, Inverse Chebyshev

3 Calculate n

4 Form $T_n(p)$ using Table 301 page 49, or Table 304 page 52, or Table 307 page 57. Add a numerator so that $T_n(0)=1$.

5 Form $t_n(p)t_n(-p) = T_n(p)T_n(-p)$

6 Form $\rho_n(p)\rho_n(-p) = 1 - t_n(p)t_n(-p)$

7 Extract $\rho_n(p)$

8 Form $\dfrac{z_{in}}{R_S} = \dfrac{1+\rho_n(p)}{1-\rho_n(p)}$ $\quad or \quad$ $\dfrac{z_{in}}{R_S} = \dfrac{1-\rho_n(p)}{1+\rho_n(p)}$

9 Perform a continued fraction expansion on $\dfrac{z_{in}}{R_S}$ to produce components.

10 Draw a schematic.

5 Filter Transient Response

The AC Spice programs in Chapter 4 are reproduced here as TRAN Spice programs. Program 40xx is renumbered as 50xx. In Figure 50611 compare the Butterworth response to the Bessel and Chebyshev responses.

Figure 50211 Butterworth Filter n=3, n=5 Transient Response

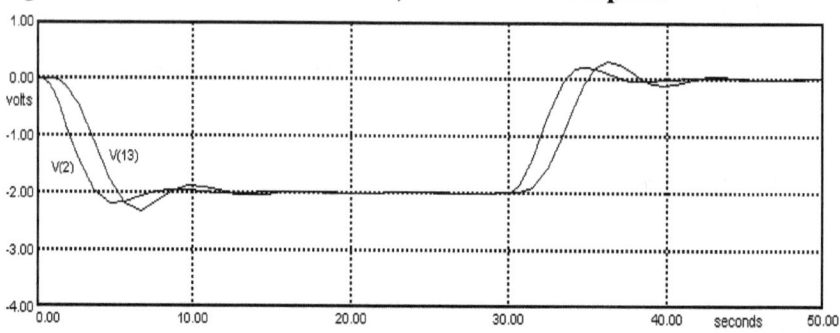

Spice program 5021 Butterworth Filter Z Synthesis

```
Fig5021.ckt  Butterworth Filters n=3,5
I1 1 0 PULSE(0 2 0 10p 10p 30 400)
C3 1 0 1.5
L2 1 2 1.333
C1 2 0 0.5
R1 2 0 1


I11 11 0 PULSE(0 2 0 10p 10p 30 400)
C15 11 0 1.545
L14 11 12 1.694
C13 12 0 1.382
L12 12 13 0.894
C11 13 0 0.309
R11 13 0 1
*-3db at omega=1 or f=1/2π          ω=2πf=2π/2π=1
*The ω=1 cutoff frequency is responsible for 50 second display.
.TRAN 0.001 50 0
.TEMP 27
.PLOT TRAN V(2) V(13) -4,1
.PRINT TRAN V(2) V(13)
.end
```

Figure 50221 Butterworth Filter n=3, n=5 Transient Response

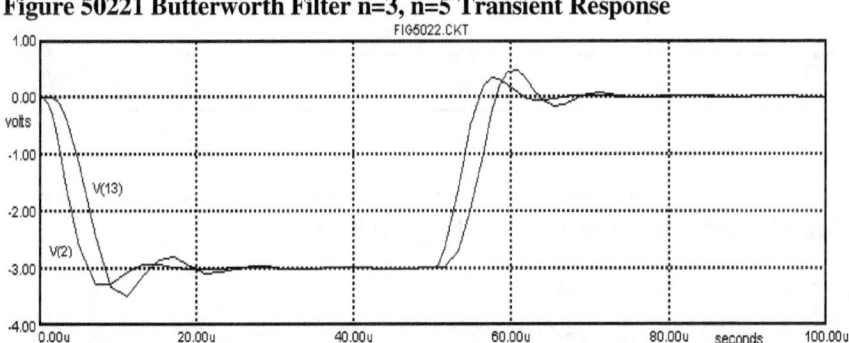

Spice program 5022 Butterworth Filter Z Synthesis

Fig5022.ckt   Butterworth Filters n=3,5

```
I1 1 0 PULSE(0 5mA 0 10p 10p 50u 20000u)
C3 1 0 3980p     ;1.5
L2 1 2 1273u     ;1.333
C1 2 0 1327p     ;0.5
R1 2 0 600       ;1


I11 11 0 PULSE(0 5mA 0 10p 10p 50u 20000u)
C15 11  0 4100p  ;1.546
L14 11 12 1619u  ;1.695
C13 12  0 3667p  ;1.381
L12 12 13  854u  ;0.894
C11 13  0  820p  ;0.309
R11 13  0  600   ;1


* -3db at f=100K hertz
.TRAN 1e-008 0.0001 0
.TEMP 27
.PLOT TRAN V(2) V(13) -4,1
.PRINT TRAN V(2) V(13)
.end
```

Figure 50311 Chebyshev Filter n=3 Transient Response

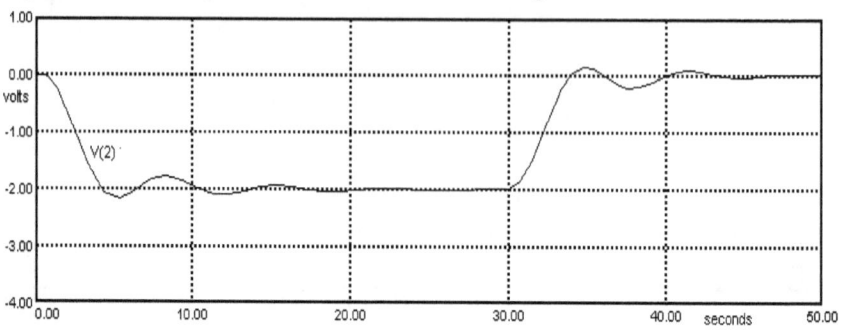

Spice Program 5031 Chebyshev Filter Z Synthesis

Fig5031.ckt   Chebyshev Filter n=3

```
I1 1 0 PULSE(0 2 0 10p 10p 30 400)
C3 1 0 1.5089
L2 1 2 1.3332
C1 2 0 1.1018
R1 2 0 1

* -3db at f=1/2pi Hertz
 TRAN 0.001 50 0
.TEMP 27
.PLOT TRAN V(2) -4,1
.PRINT TRAN V(2)
.end
```

Figure 50321 Bessel Filter n=3 Transient Response

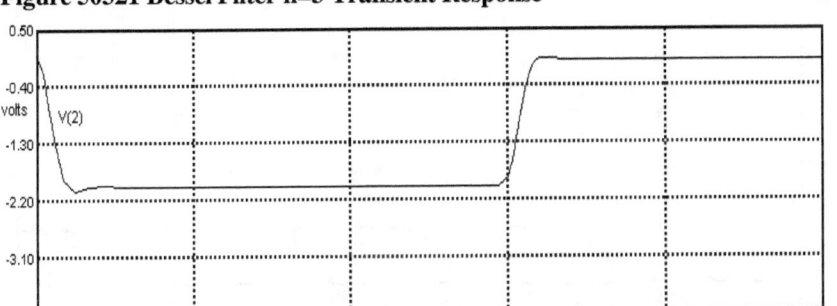

Spice Program 50321 Bessel Filter Z Synthesis

Fig5032.ckt   Bessel Filter n=3

```
I1 1 0 PULSE(0 2 0 10p 10p 30 400)
C3 1 0 0.83333    ; Figure 403
L2 1 2 0.4800
C1 2 0 0.1667
R1 2 0 1
```

```
* -3db at f=1/2pi Hertz
.TRAN 0.001 50 0
.TEMP 27
.PLOT TRAN V(2) -4,0.5
.PRINT TRAN V(2)
.end
```

Analog Filter Design

Figure 50611 Three Filters n=5 Transient Response

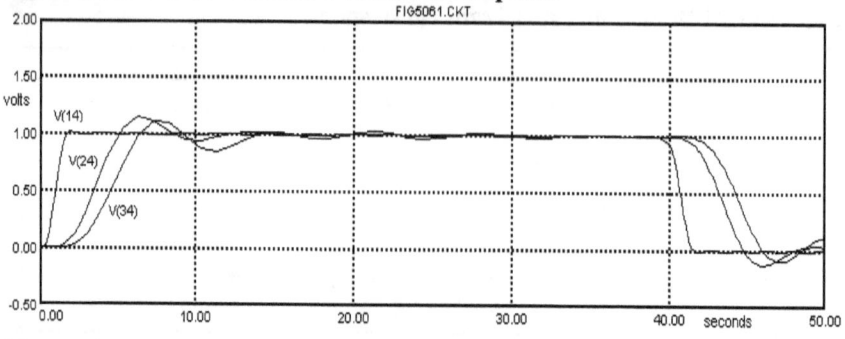

Spice Program 5061 Three Filters Darlington Synthesis

```
Fig5061.ckt    Filters n=5
V11 11 0  PULSE(0 2 0 10p 10p 40 400)
R10 11 12 1    ;Butterworth
C11 12  0 0.618
L12 12 13 1.618
C13 13  0 2
L14 13 14 1.618
C15 14  0 0.618
R16 14  0 1

V21 21 0  PULSE(0 2 0 10p 10p 40 400)
R20 21 22 1    ;Chebyshev
C21 22  0 2.135
L22 22 23 1.091
C23 23  0 3.001
L24 23 24 1.091
C25 24  0 2.135
R26 24  0 1

V31 31 0 PULSE(0 2 0 10p 10p 40 400)
R30 31 32 1    ;Bessel
C31 32  0 0.072
L32 32 33 0.209
C33 33  0 0.331
L34 33 34 0.458
C35 34  0 0.930
R36 34  0 1

.TRAN 0.001 50 0
.TEMP 27
.PLOT TRAN V(14) V(24) V(34) -0.5,2
.PRINT TRAN V(14) V(24) V(34)
.end
```

Figure 50621 Butterworth Filters n=5 Transient Response

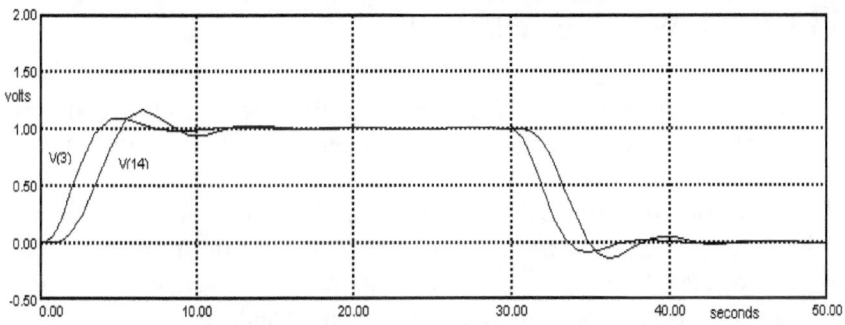

Spice Program 5062 Butterworth Filters Darlington Synthesis

```
Fig5062.ckt    Filters n=3,5
V1 1 0 PULSE(0 2 0 10p 10p 30 400)
R1 1 2 1        ; figure 405
C1 2 0 1
L1 2 3 2
C2 3 0 1
R2 3 0 1


R11 1  12 1
C15 12  0 0.618 ; figure 406
L14 12 13 1.618
C13 13  0 2.000
L12 13 14 1.618
C11 14  0 0.618
R12 14  0 1

.TRAN 0.001 50 0
.TEMP 27
.PLOT TRAN V(3) V(14) -0.5,2
.PRINT TRAN V(3) V(14)
.end
```

6 How to Write Spice Programs

A Spice program requires a *title statement* as the first line, a (dot end) *.end statement* as the last line, and a *.temp* statement that specifies temperature.

Required program lines Between the first and last lines you insert, in any order, a list of *data statements* that describe the components of the circuit to be simulated, and a list of *control statements* that describe the circuit analysis to be performed. Any line can be empty/blank.

Comments An asterisk (*) in the first column indicates that the line is a comment line. A semicolon (;) anywhere in a line means the rest of the line is a comment. Comment lines may be placed anywhere in a Spice program.

Fig2011.ckt   *;title statement must be on the first line*

*The R3 line is a data statement

R3 5 7 8.2K   *;8.2K resistor connected to nodes 5 and 7*
.TEMP 27
.end   *; dot end must be on last line, ends the program*

Our numbering scheme   We write a Spice program on a word processor such as Notebook or Wordpad, when we want to evaluate the performance of the circuit in Figure 402 for example. We save the text as the text file Fig4021.ckt (all Spice programs are text files). However see the note below. The first line of a program includes the name Fig4021.ckt. Plots of results from program Fig4021.ckt are labeled Fig40211, Fig40212, etc. A second program for Figure 402 is given the name Fig4022.ckt. Plots are labeled Fig40221, Fig40222, and so forth. In this way we know how circuit figures, Spice programs, Spice files, and plots are related.

> *A Spice program is a text file created on any word processor.*

Note: We only use the spice-text feature of the commercial Spice programs that are really complex wrap arounds to the basic Berkeley Spice. Click on File, New and select spice-text. Type into the blank screen page (your Spice may be different).

6.1 AC Spice program Fig4021.ckt

Fig4021.ckt   factor 1/(p+a)

First line The first line of every program is assumed by Spice to be a title statement. The title statement can include any words.

V1 1 0 AC 1 0   ; volts

Voltage source Signal generator v_1 is connected to nodes 1 and 0. The v_1 signal is a AC sinewave with 1 volt magnitude and 0 phase. V1 is defined as a one volt sinewave for *all* frequencies

R1 1 2 1000
C1 2 0 .0159155u

Circuit components Resistor R_1 is connected to nodes 1 and 2. Capacitor C_1 is connected to nodes 2 and 0 (Figure 402)

.AC DEC 200 10 1e+007

AC control statement Dot AC DEC defines a frequency range from 10 Hz to 10^7 Hz with points to be calculated every 200Hz.

.PLOT AC VDB(2) -40,10   ;magnitude dB
*.PLOT AC VP(2) -100,0   ;phase degrees
.PRINT AC V(2)   ;produces numeric output data
.PRINT AC VDB(2)

Plot the data dot plot AC executes the dot AC statement calculating the v_2 magnitude in dB, or phase in degrees, from 10Hz to 10^7Hz. The asterisk that defines a line as a comment deactivates the VP(2) phase plot. Dot print records numeric data such as dB vs frequency.

.TEMP 27

Temperature Dot temp (.temp) defines temperature as 27 degrees C.

.end

Last line The last line of every program is .end (dot end).

V_1 is the Spice voltage source connected from node 1 to node 0, and AC 1 0 means the source is sinusoidal with 1volt peak amplitude and 0 degrees phase. The dot AC (.AC) line defines a frequency range of 10Hz to 10MHz incremented by 200Hz.

Spice starts by selecting 10Hz, and calculating the v_2/v_1 ratio (i.e. $|T(j2\pi10)|$). Then Spice increments the frequency by 200 Hz to 210Hz, and calculates $|T(j2\pi\,210)|$. The process is repeated every 200Hz up to 10MHz. The dot plot lines produce Figures 40211 and 40212.

Analog Filter Design

The corner frequency is at 10KHz. The magnitude decreases 20dB over the decade 100KHz to 1MHz. The slope is −20dB/decade, −6dB/octave. The phase at 10KHz corner is −45 degrees.

Spice Program 4021 Calculates Frequency Response

Fig4021.ckt   factor 1/(p+a)
V1 1 0 AC 1 0   ; volts
R1 1 2 1000
C1 2 0 .0159155u
*.PLOT AC VDB(2) -40,10 ;magnitude dB
.AC DEC 200 10 1e+007
.PLOT AC VP(2) -100,0   ;phase degrees
.PRINT AC V(2)
.PRINT AC VDB(2)
.TEMP 27
.end

Figure 402

Figure 40211 Magnitude of T(p)=V$_2$/V$_1$

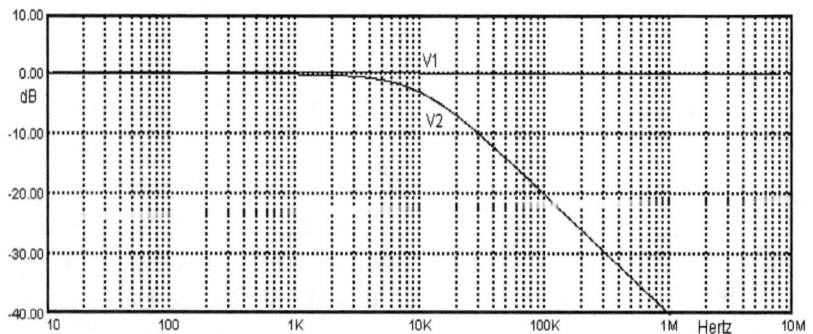

Figure 40212 Phase of T(p) =V$_2$/V$_1$

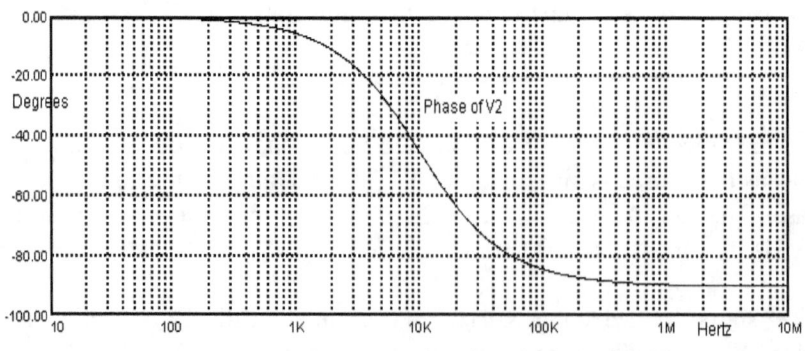

6.2 TRAN Spice program Fig5031.ckt

Fig5031.ckt   Chebyshev Filter n=3

First line The first line of every program is assumed by Spice to be a title statement. The title statement can include any words.

```
I1 1 0 PULSE(0   2    0    10    10    30     400)
```
I_{lo} I_{hi} t_{delay} t_{rise} t_{fall} t_{width} t_{period}

Voltage source Pulse generator I_1 is connected to nodes 1 and 0. The pulse is a step function in effect, because width is only 30s and the 400s period greatly exceed the 50s plot time.

```
C3 1 0 1.5089
L2 1 2 1.3332
C1 2 0 1.1018
R1 2 0 1
```

Circuit components Capacitor C_3 is connected to nodes 1 and 0. Inductor L_1 is connected to nodes 1 and 2. Capacitor C_1 is connected to nodes 2 and 0. Resistor R_1 is connected to nodes 2 and 0.

```
.TRAN 0.001 50 0
```

TRAN control statement Dot TRAN defines a time range from 0 to 50s that is incremented every 1ms (`1e-003, 0.001`).

```
.PLOT TRAN V(2) -4,1
.PRINT TRAN V(2)   ;produces numeric output data
```

Plot the data dot plot TRAN executes the dot TRAN statement calculating the v_2 waveform from 0s to 50s every 1ms.

```
.TEMP 27
```

Temperature Dot temp (.temp) defines temperature as 27 degrees C.

```
.end
```

Last line The last line of every program is .end (dot end).

Analog Filter Design

Figure 50311 Chebyshev Filter n=3 Transient Response

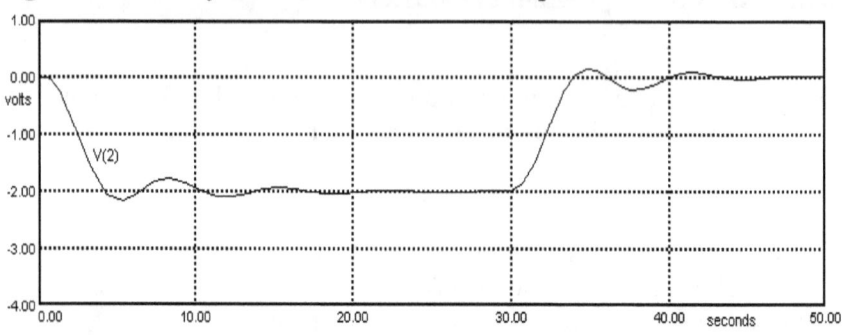

Spice Program 5031 Chebyshev Z Synthesis

Fig5031.ckt   Chebyshev Filter n=3

```
I1 1 0 PULSE(0 2 0 10p 10p 30 400)
C3 1 0 1.5089
L2 1 2 1.3332
C1 2 0 1.1018
R1 2 0 1

.TRAN 0.001 50 0
.TEMP 27
.PLOT TRAN V(2)  4,1
.PRINT TRAN V(2)
.end
```

Eight Experiments

The experiments support the text. You teach yourself how to use electronic instruments and tools, how to make measurements, and learn about parts. You teach yourself how to build the circuits shown in the figures on a solderless breadboard, and measure their performance.

Electrical engineering book learning is necessary but not sufficient, because an EE has to know

1 how to make measurements in order to evaluate a design.
2 what parts are available as well as their properties.
3 what parts look like, and how to read any part's label.
4 the equivalent circuit of a part given a frequency range,
5 that all parts have parasitic components attached to them.
6 what parasitic components are added when a part is placed in a circuit.
7 how to use the tools of the trade
8 and so on.

Here you design and build circuits from the get go, while referring to the text and other writings to learn what you need to know to implement the design. The idea is that you seek answers on an ongoing basis to the many questions that arise as you try to design and build circuits.

You take time outs to seek answers to the questions by reading the text, doing the problems, and perhaps doing a search on the Internet. In this way theory and practice merge.

Parts

A comprehensive view of available parts is found in the Product Index on the web site of any electronics parts distributer. The Electronic Circuits experiments only use a small subset of each type of part such as resistors R, capacitors C, inductors L, transformers, potentiometers R, and transistors.

Associated with every part is a *data sheet*, which presents the part's characteristics such as text descriptions, thermal characteristics, tables of electrical characteristics, available electrical values, package dimensions, pin assignments, test circuits, application notes, and so forth.

A *data sheet* tells you what a part is about.

Rarely will you find in a *data sheet* the equivalent circuit of a part given a frequency range, nor any information about the part's parasitic components attached to them. This information is usually found in *technical articles* or *application notes* issued by the manufacturer.

Every manufacturer has a web site from which you can download data sheets, Spice models, application notes, white papers and so forth.

Each experiment specifies the parts used in the experiment's circuits.

Instruments

What do you need to measure or observe, while evaluating circuit performance? As a minimum you need to measure or observe ohms, DC voltages, DC currents, steady state AC signal voltages, and transient state signal voltages.

You need an oscilloscope so that you can see the DC voltages and AC signal voltages at circuit nodes. Two channels allow you to compare what is going on at two nodes, such as an input and its corresponding output. For our purposes here a two channel oscilloscope, with 1MHz bandwidth or better, is satisfactory.

You need a signal generator, specifically a function generator, to generate the signals driving the circuit under test. Function generator, because you will need sine waves, square waves, and pulses. A function generator with maximum frequency 1MHz or better will do.

A very important feature of any signal generator is that a signal amplitude does NOT change when frequency is changed. This must be verified as experiments are executed.

You need a DC multimeter, which measures wide ranges of volts, ohms, and amperes. Analog or digital meter? Your choice. We use both types.

To us a Power Supply is a *generator* of zero frequency DC voltages. In this text's experiments you need ±5V linear power supply.

Tools

Side cutters and long nose pliers cut and form wire. A wire stripper strips insulation from the ends of wire without damaging the wire. A lead forming tool accurately forms wire leads of parts for insertion into the breadboard holes. Tweezers facilitate picking up and placing small parts. Clip leads connect terminals as needed.

1	reactance chart (do a Google search for reactance chart)
1	breadboard
1	side cutters, 4"
1	long nose pliers, 4"
1	wire stripper
1	lead forming tool
1	tweezers, fine point
?	Clip leads

Color Code

An effective way to learn the color code is to sort a pile of resistors by value. Another way is to use the program *colorcode.exe*.

1st band	1st digit
2nd band	2nd digit
3rd band	number of zeros
4th band	tolerance, gold 5%, silver 10%

digit	0	1	2	3	4	5	6	7	8	9
color	black	brown	red	orange	yellow	green	blue	violet	gray	white
10% values	100	120	150	180	220	270	330	390	470	560	680	820	1000

5% values

100	110	120	130	150	160	180	200	220	240	270	300	330
360	390	430	470	510	560	620	680	750	820	910	1000	

Examples

22	red, red, black
220K	red, red, yellow
1.2K	brown, red, red
47K	yellow, violet, orange
910	white, brown, brown
8.2M	gray, red, green

Contents

Safety issue - The AC Line Voltage is extremely dangerous

SAFETY FIRST!!!!!!!

SAFETY FIRST!!!!!!!   Electricity is silent so be very careful. We remove all metal objects from our hands and wrists such as rings and watches. If you do not remove them, then know you are taking an unnecessary risk.

Furthermore you do the experiments at your own risk, because there is no way we can supervise your work.

Word to the wise - NEVER GRAB anything, because if it's hot you will have difficulty letting go.

The AC Line Voltage is extremely dangerous.

The Solderless Breadboard and the Power Supply

Solderless breadboards are designed to connect parts together without using solder. A power supply energizes the circuits on the solderless breadboard.

You build a circuit by inserting leads in holes. For example a resistor has 2 leads, which when suitably bent can be inserted into 2 pin holes. There is no need to shorten a part's leads in these experiments.

A part's lead (28 to 20 AWG, 0.0126 to 0.0320 inches diameter) is inserted into a hole whose spring loaded metal insert grabs the lead. The metal inserts are not visible. *The metal inserts in horizontal rows of 5 holes are shorted together.* Consequently leads inserted in the same 5-hole-row are shorted together. The leads placed in a 5 hole row are the equivalent of leads soldered together. A row is equivalent to a circuit node.

Figure 101 Part of a Solderless breadboard showing fields of holes.

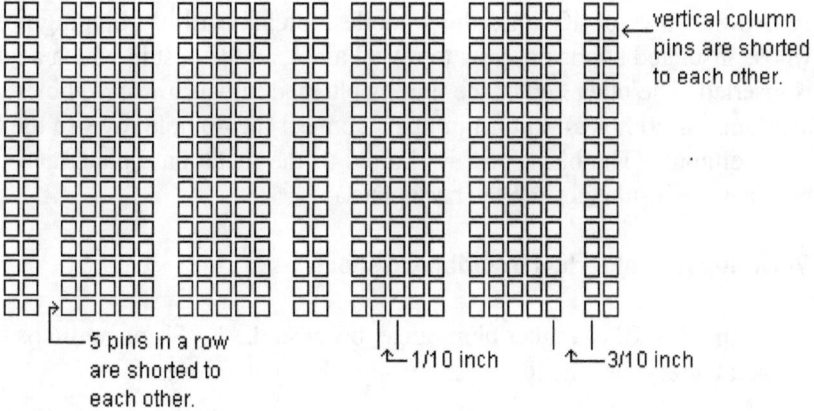

vertical column pins are shorted to each other.

5 pins in a row are shorted to each other.

1/10 inch   3/10 inch

A board is an assembly of *two types of hole patterns* mounted on a metal plate (Figure 101). One type of hole pattern has two 5 × 59 arrays of holes separated by a narrow gutter. The columns of rows parallel and adjacent to the gutter are 0.3 inches apart, because 0.3 inches is the IC DIP package minimum pin row spacing. Each array of holes is a column of 59 five hole rows. All holes are on a 0.1 inch grid so that vertical and horizontal separation is 0.1 inch. The five holes in each row are shorted together. The rows are not shorted together.

IC pins are inserted into the board so that the IC straddles the gutter and each IC pin plugs into one hole of one row. Then the four other holes in each pin row are available to receive leads, which are automatically connected to the IC pin. In this way a circuit is wired from node to node. Larger ICs have 0.4-inch and 0.6-inch pin row spacing. These are inserted in the same way that the narrower 0.3 inch ICs are inserted; however, the covered up row holes are not available for point-to-point wiring.

The 3 leads of a discrete transistor are inserted in 3 *different* rows.

The 3 leads of a potentiometer are inserted in 3 *different* rows

Power and Ground The other type of hole pattern is formatted to distribute power and ground. The pattern has two columns of 50 holes. The 50 holes *in each column* are shorted together. The two columns are not shorted together so that one column can distribute 5 volts, for example, and the other 0 volts (ground). Another pattern across the top of the board has two rows of 40 holes. The 40 holes *in each row* are shorted together.

Above the rows of 40 holes the solderless breadboard has binding posts whose insulated knobs unscrew to reveal a hole in the post in which a wire is inserted. The other end of the wire is plugged into a row of 40 holes (or a column of 50 holes). In turn jumpers connect the 40 hole rows to the 50 hole columns. (The black post is shorted to the metal base plate, and the red ones are insulated from the base plate.)

Verifying the solderless breadboard shorts

Select the R × 10 or higher ohm range, because the R × 1 range drains the 1.5 volt battery (use the R × 1 range only when you have to).

Connect a pin to each lead of your ohmmeter. Place one pin in the first hole of a 5 hole row. Place the other pin in each of the other 4 holes in the row. Verify that the resistance is essentially zeros ohms (a short).

Repeat for the columns of holes.

Measure the resistance between rows, which should be infinite (open circuit). Repeat for row to columns, etc. Check out all possibilities to KNOW how the holes are wired.

Power Supply We use a linear open frame $\pm 5V$ power supply. We have to add a power cord. COVER THE AC TERMINALS WITH TAPE. Plug the power cord into a plug strip outlet. The plug strip on/off switch becomes the power supply on/off switch. Think of a power supply as a constantly recharged battery.

The source impedance R_S of the battery is estimated as follows. If a 5V output drops by 2%, or 0.1V, when a 0.25 ampere load current is drawn, then $R_S=0.1V/0.25A=0.4$ ohms. However if plus is shorted to minus, then potentially $I=5/0.4=12.5$ amperes, which may or may not flow. This is why unprotected parts are destroyed (and perhaps the power supply).

We have to be very careful to avoid shorting plus to minus.

> *Shut off the power supply OR Disconnect the voltage leads AT THE SUPPLY* when making changes on the solderless breadboard.

Connecting the power supply to the solderless breadboard

Connect the binding posts to the solderless breadboard. The solderless breadboard has binding posts whose insulated knobs unscrew to reveal a hole in the post. The black post is shorted to the metal base plate, and the red ones are insulated from the base plate. Unscrew the black post and insert a black wire in the post hole. Tighten the black knob to secure the wire. Insert the other end of the wire into a hole in a column of holes that you want to be grounded. Repeat with a red wire from a red post to what becomes the power supply voltage, the B+ column of holes (e.g. +5V). Repeat for B− (e.g. −5V).

Connect the power supply to the binding posts First turn off the AC power to the supply. Use clip leads or wires to connect the power supply voltage terminals to the solderless breadboard binding posts.

AFTER you have built a circuit, turn on the AC power to the supply.

Turn the power off *before* you make circuit changes.

Use the multimeter to verify the $\pm 5V$ voltages.

AC Voltmeter

Build the AC Voltmeter circuit (Figure 201) at the right side of the breadboard. You will use it in experiments. The meter is a 1mA full scale DC current meter. We used the Texas Instruments BiMOS quad op amp **TLC074CN**. Any comparable MOS op amp will do. The diodes are 1N914. Any low current diodes will do. R_1 is 10^6 ohms, R_2 is 1K ohms.

Make sure all connections to the breadboard are zero ohms. This BiMOS op amp has a parasitic 20pF capacitor at its inputs ($10^6\Omega$ at 8KHz, $10^3\Omega$ at 800KHz).

Connect a sine wave signal generator as V_{in}. Adjust the signal's frequency to about 1KHz. Adjust the signal's amplitude until the meter reads 1mA. The peak to peak amplitude should be 3.14 volts (1.57V peak).

Use an oscilloscope to verify that the signal at pins 5,6,7,2,3 is an undistorted sine wave. To check meter bandwidth increase the frequency until the meter reading drops to 0.89mA (−1dB). When we did that the frequency was 800KHz. *Important - See Experiment 8 about the 2700μμF capacitor.*

$$for\ sinewave\ I_{max}\sin\omega t \quad I_{average}=\frac{2}{\pi}I_{max} \Rightarrow I_{max}=\frac{\pi}{2}I_{average}$$

$$V_{2max}=V_{3max}=V_{7max}=V_{5max}=V_{inmax}$$
$$V_{inmax}=1.57volts\ for\ full\ scale\ 1mA\ meter\ dc\ current$$

Figure 201 AC Voltmeter 1MHz

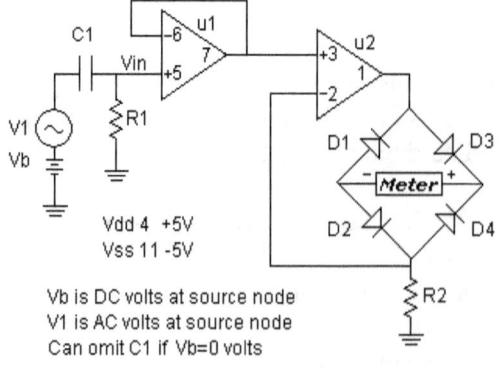

Figure 202 TI TLC074CN

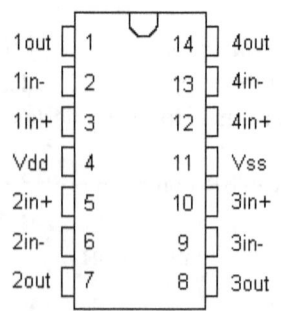

There is a lot going on here. I.e. an opportunity to teach yourself about the measuring instruments.

Connect meter + to D3-D4 junction, meter − to D1-D2 junction

> **The AC voltmeter input is at 0V DC. When measuring a signal at a node that is NOT at zero volts DC connect a capacitor (e.g. $C_1=1\mu F$) from that node (C_1+) to the AC voltmeter input (C_1−).**

Figure 2011 AC Voltmeter Circuit

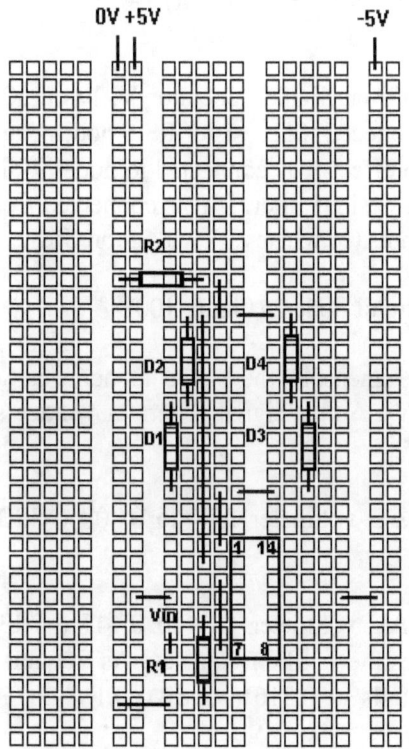

> Note for experienced oscilloscope users. The oscilloscope can be used to measure dB and period T=1/frequency f.

Analog Filter Design

Experiment 1 Elementary RLC Filters

The purpose here is to show that 4 types of transfer functions can be implemented with 1 each R, L, and C.

One each R, L, and C (R=various Ω, L=1000μH, C=2700pF) can be assembled to form low pass, high pass, band pass, and band reject filters. Standard form parameters are

$$(1a) \quad Q_P = \frac{R}{\omega_0 L} \qquad (1b) \quad \omega_0^2 = \frac{1}{LC} \qquad (1c) \quad \lambda = \frac{p}{\omega_0} \qquad (1d) \quad T(p) = \frac{v_2}{v_1}$$

Experimental procedure for any Filter Signal generator V_1 has a 50Ω source resistance R_S. For example connect 510Ω resistor R_a to nodes 3 and 2 to make R= 560Ω (Figure 100). In effect R is connected to nodes 1 and 2. Connect L and C according to the desired filter. If R is connected from node 2 to ground (node 0) use 560 ohms (or whatever), and ignore R_S.

Connect the AC voltmeter input to circuit node 2 (Figure 100).

For example - for a low pass filter measure the signal at node 2. Set frequency to about 1KHz. Monitor V_2. Adjust signal generator voltage so that AC voltmeter reads 1 (0dB).

Increase frequency gradually until AC voltmeter falls to 0.707 (−3dB). Record the frequency.

Increase frequency gradually until AC voltmeter falls to 0.447 (−7dB). Record the frequency. This is 2 times the −3dB frequency. As frequency is increased AC voltmeter reading falls to 0.316 (−10dB), 0.1 (−20dB), and so forth.

Figure 101 Sample Layout

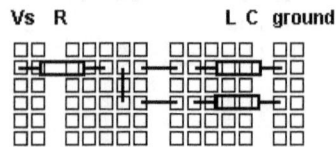

Figure 100 Sample Circuit

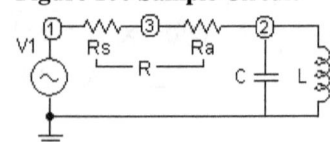

1 A series L and shunt C produce a

−12dB/octave slope above ω_0, because $z_L \to \infty$ and $z_C \to 0$ as $\omega \to \infty$ (Figure x101). In other words there are two transmission poles, because $T(\lambda) \to 0$ as $1/\lambda^2$ as $\lambda \to \infty$. Since $T_{LP}(\lambda)=1$ from 0 to ω_0 this circuit is a low pass filter (Figure x10111). Calculate Q.

Figure x101 Low Pass

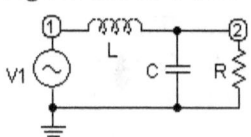

(2) $$T_{LP}(p) = \frac{y_L}{y_L + y_{RC}} = \frac{1/pL}{1/pL + g + pC} = \frac{1}{1 + gLp + LCp^2}$$

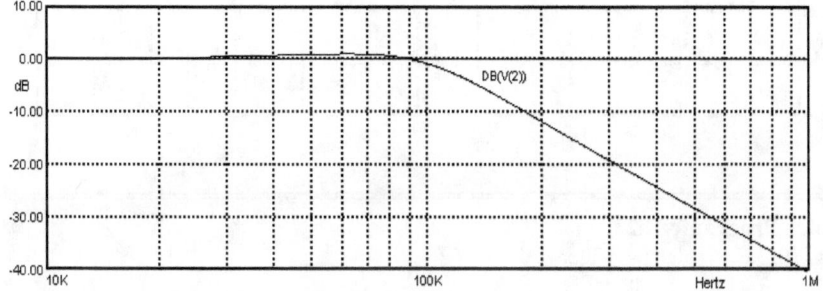

$$= \frac{1}{1 + \dfrac{L}{R}\dfrac{\omega_0}{\omega_0}p + LCp^2\dfrac{\omega_0^2}{\omega_0^2}} = \frac{1}{1 + \dfrac{1}{Q_p}\dfrac{p}{\omega_0} + \dfrac{p^2}{\omega_0^2}} = \frac{1}{1 + \dfrac{1}{Q_P}\lambda + \lambda^2}$$

Figure x10111 Low Pass Transmission Magnitude of T(p)

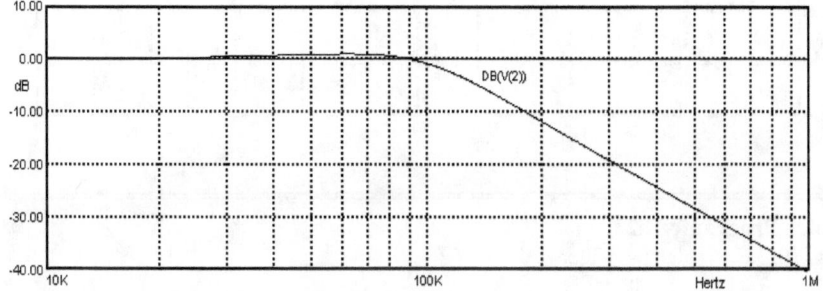

Spice Program x1011

```
Figx1011.ckt    low pass filter
V1 1 0 AC 1 0   ; volts
*names/nodes/values
L 1 2 1000u
C 2 0 2700p
R 2 0 560
*.PLOT AC VP(1) VP(2) -200,0
.AC DEC 400 10000 1e+006
.TEMP 27
.PLOT AC VDB(1) VDB(2) -40,10
.end
```

103

2 Derive equation 3. Design a high pass filter circuit (Fig x102) with −3dB at about 100KHz, $z_0 = 560\Omega$, output is v_2. Find R, L, C, ω_{-3dB}, Q. Write Spice program x1021 that produces a plot of $T_{HP}(\omega)$.

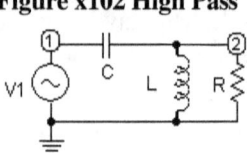

Figure x102 High Pass

(3) $\quad T_{HP}(\lambda) = \dfrac{\lambda^2}{1 + \dfrac{1}{Q_P}\lambda + \lambda^2}$

Figure x10211 High Pass Transmission Magnitude of T(p)

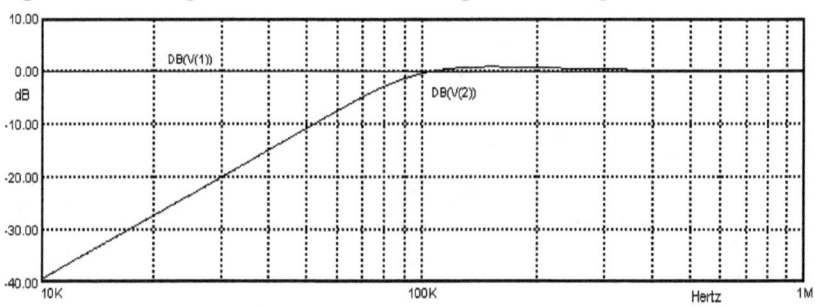

═══

Spice Program x1021

```
Figx1021.ckt    high pass filter
V1 1 0 AC 1 0     ; volts
*names/nodes/values
C 1 2 2700p
L 2 0 1000u
R 2 0 560
*.PLOT AC VP(1) VP(2) 0,200
.AC DEC 400 10000 1e+006
.TEMP 27
.PLOT AC VDB(1) VDB(2) -40,10
.end
```

═══

3 Derive equation 4. Write Spice program x1031 that plots $T_{BP}(\omega)$. R=56000Ω. Find Q. Observe that R is in parallel with LC.

Figure x103 Band Pass

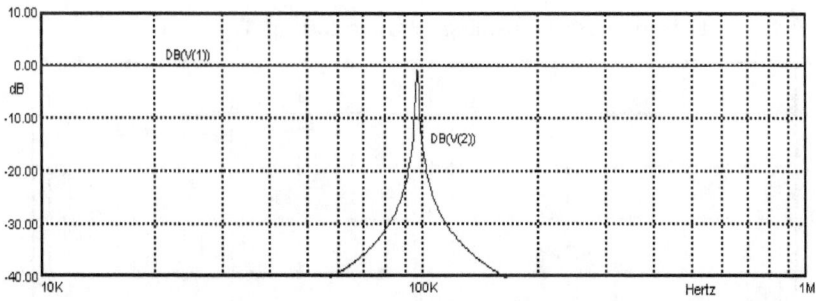

$$(4) \quad T_{BP}(\lambda) = \frac{1}{Q_P} \frac{\lambda}{1 + \dfrac{1}{Q_P}\lambda + \lambda^2}$$

Figure x10311 Band Pass Transmission Magnitude of T(p)

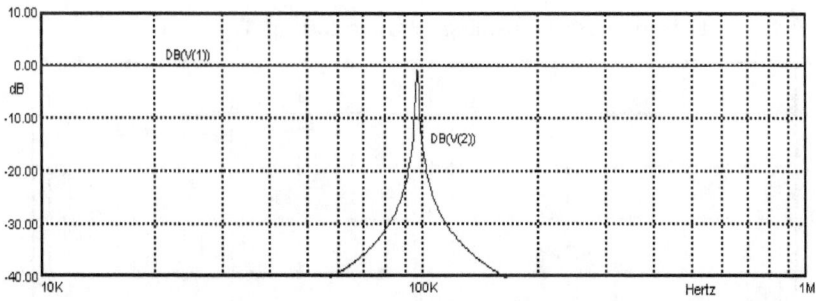

Spice Program x1031
```
Figx1013.ckt    band pass filter
V1 1 0  AC 1  0   ; volts
*names/nodes/values
R 1 2 56000
C 2 0 2700p
L 2 0 1000u
*.PLOT AC VP(1) VP(2) 0,200
.AC DEC 400 10000 1e+006
.TEMP 27
.PLOT AC VDB(1) VDB(2) -40,10
.end
```

---

4 Derive equation 5. Design a band reject circuit bandwidth=10KHz, f_0=100KHz. Find R, L, C, ω_1, ω_2, Q. Write Spice program x1041 that produces a plot of $T_{BR}(\omega)$.

Figure x104 Band Reject

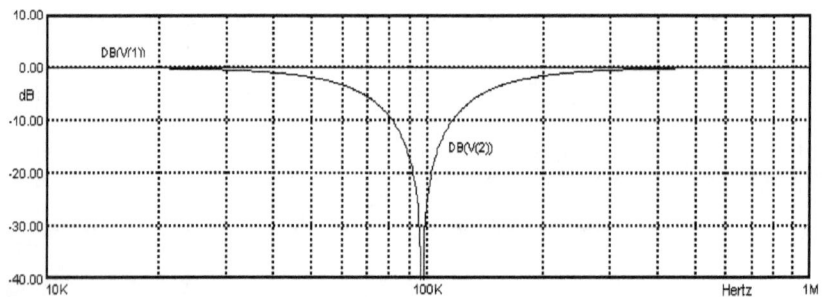

(5) $\quad T_{BR}(\lambda) = \dfrac{1+\lambda^2}{1+\dfrac{1}{Q_P}\lambda+\lambda^2}$

Figure x10411 Band Pass Transmission Magnitude of T(p)

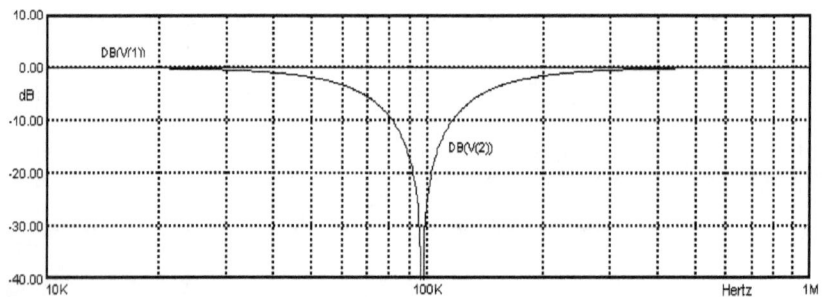

Spice Program x1041
```
Figx1041.ckt    band reject filter
V1 1 0 AC 1 0   ; volts
*names/nodes/values
C 1 2 2700p
L 1 2 1000u
R 2 0 560
*.PLOT AC VP(1) VP(2) 0,200
.AC DEC 400 10000 1e+006
.TEMP 27
.PLOT AC VDB(1) VDB(2) -40,10
.end
```

Experiment 2 Design of Ladder Filters

Design a constant k section and an m derived section when R=50Ω, f_C=50/π KHz, and m=0.6 (pages 27, 30, 42, Spice program x2011).

Change the exact values from the design to standard values available in industry (Spice program x2021).

Compare the results (Figures x20111, x20211. The f infinity frequency did not change significantly (from 20KHZ to 19.9KHz). However the maximum attenuation decreased from 121dB to 69dB, which is still very significant.

Figure x20111 Low Pass Transmission Exact Values for T(p)

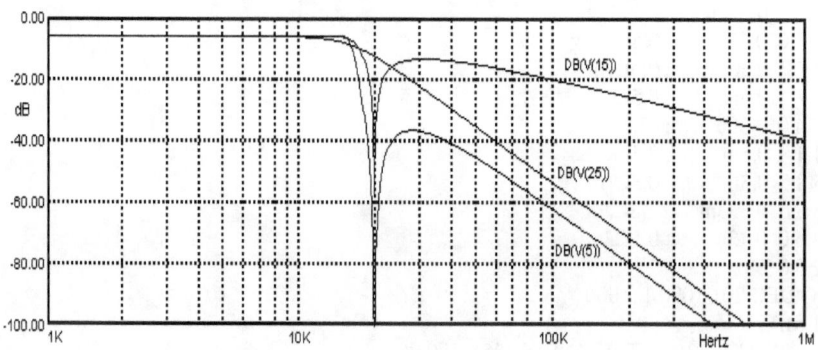

Figure x20211 Low Pass Transmission Standard Values for T(p)

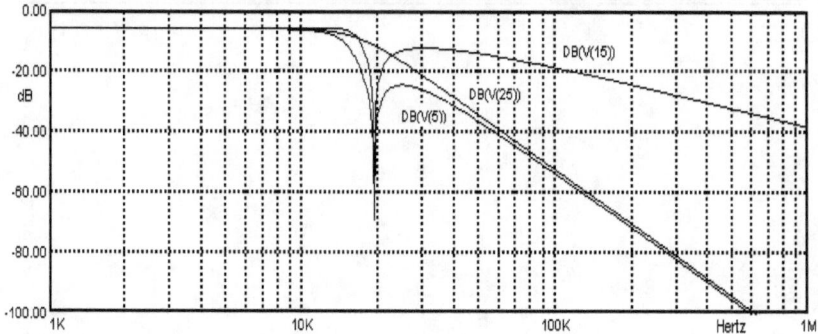

Spice Program x2011
```
Figx2011.ckt  Low pass filters
V1 1 0 AC 1 0
R21  1 22 50       ; constant k
L22k 22 24 0.5m    ; 0.5Lk
C22k 24  0 0.39u   ; Ck
L23k 24 25 0.5m    ; 0.5Lk
R22  25 0 50


R11  1 12  50   ; m derived
L11m 12 13  0.3m      ; 0.5L1m
L12m 13 14  0.533m ; 2L2m
C12m 14 0  0.12u    ; 0.5C2m
L13m 13 15  0.3m     ; 0.5L1m
R12  15 0  50


R1 1 2  50        ; cascade k and m
L1 2 3  0.533m     ; 2L2m
C1 3 0  0.12u     ; 0.5C2m
L2 2 7  0.3m      ; 0.5L1m
L3 7 4  0.5m      ; 0.5Lk
C2 4 0  0.39u     ; Ck
L4 4 8  0.5m      ; 0.5Lk
L5 8 5  0.3m      ; 0.5L1m
L6 5 6  0.533m     ; 2L2m
C4 6 0  0.12u     ; 0.5C2m
R2 5 0  50
.AC DEC 201 1000 1e+006
.TEMP 27
.PLOT AC VDB(5) VDB(15) VDB(25) -100,0
.PRINT AC VDB(5) VDB(15) VDB(25)
.end
```

Spice Program x2021

```
Figx2021.ckt  Low pass filter
V1 1 0 AC 1 0
R21   1 22 50        ; constant k
L22k 22 24 0.47m     ; 0.5Lk
C22k 24  0 0.39u     ; Ck
L23k 24 25 0.47m     ; 0.5Lk
R22  25  0 50

R11   1 12  50       ; m derived
L11m 12 13  0.27m    ; 0.5L1m
L12m 13 14  0.56m    ; 2L2m
C12m 14  0  0.12u    ; 0.5C2m
L13m 13 15  0.27m    ; 0.5L1m
R12  15  0  50

R1 1 2  50           ; cascade k and m
L1 2 3  0.56m        ; 2L2m
C1 3 0  0.12u        ; 0.5C2m
L2 2 7  0.27m        ; 0.5L1m
L3 7 4  0.47m        ; 0.5Lk
C2 4 0  0.18u        ; 0.5Ck
L4 4 8  0.47m        ; 0.5Lk
L5 8 5  0.27m        ; 0.5L1m
L6 5 6  0.56m        ; 2L2m
C4 6 0  0.12u        ; 0.5C2m
R2 5 0  50

.AC DEC 201 1000 1e+006
.TEMP 27
.PLOT AC VDB(5) VDB(15) VDB(25) -100,0
.PRINT AC VDB(5) VDB(15) VDB(25)
.end
```

Experiment 3 Design an Op Amp Low Pass Filter

The non-inverting op amp circuit relates V_3 to V_1.

(1) $\dfrac{V_3}{V_1} = 1 + \dfrac{R_2}{R_1} = K$

Straightforward circuit analysis produces the transfer function.

(2a) $\dfrac{V_3}{V_6} = K \dfrac{1}{R_3 C_3 R_4 C_4} \times \dfrac{1}{p^2 + (1/R_4 C_4 + 1/R_3 C_4 + (1-K)/R_3 C_3)p + 1/R_3 C_3 R_4 C_4}$

(2b) $\dfrac{V_3}{V_6} = K\omega_3\omega_4 \times \dfrac{1}{p^2 + (\omega_4 + \omega_{34} + (1-K)\omega_3)p + \omega_3\omega_4}$

(2c) $\dfrac{V_3}{V_6} = K{\omega_0}^2 \times \dfrac{1}{p^2 + \dfrac{1}{Q}p + {\omega_0}^2}$

(3a) $\dfrac{1}{Q} = (\omega_4 + \omega_{34} + (1-K)\omega_3)$ ${\omega_0}^2 = \omega_3\omega_4$

(3b) $R_4 = R_3 = 1K\Omega$, $C_4 = C_3 = 0.01\mu F$ $\rightarrow$ $\omega_3 = \omega_4 = \omega_0$

(3c) $\dfrac{1}{Q} = (2\omega_4 + (1-K)\omega_4) = \omega_0(3-K)$ $\rightarrow$ $K = 2$ $\rightarrow$ $R_1 = R_2 = 10K$

(3d) $\dfrac{1}{Q} = \omega_0$ $\rightarrow$ (2c) $\dfrac{V_3}{V_6} - 2{\omega_0}^2 \times \dfrac{1}{p^2 + \omega_0 p + {\omega_0}^2}$

A literature survey reveals many design processes. A simplified Sallen-Key filter design process with a Butterworth response is as follows.

Figure X301 Sallen-Key Low Pass Filter

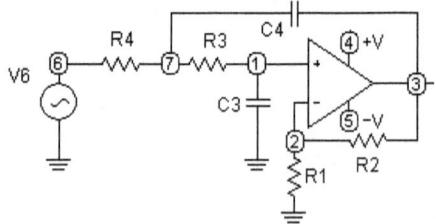

(4a) $C_4 = C_3 = 0.01\mu F$

(4b) $R_4 = R_3 = \dfrac{1}{2\pi f_0 C_3} = \dfrac{1}{2\pi \cdot (50/\pi) \cdot 10^3 \cdot 0.01 \cdot 10^{-6}} = 10^3 \,\Omega$

Build the filter, measure its performance, and write a Spice program to confirm performance.

Spice Program x3011

Figx3011.ckt Sallen-Key Low pass Filter

```
VP  4  0  DC  5
VN  5  0  DC -5
V1  6  0  AC 1
.lib G:\!B\NLPbooks\ee109pdf\ecexspice\a_tlc07.txt

*.SUBCKT tlc07a
* Node assignments
*       |non-inverting input
*       |   |inverting input
*       |   |   |positive supply
*       |   |   |   |negative supply
*       |   |   |   |   output
*       |   |   |   |   |
*       1   2   4   5   3
*X1  0  2  4  5  3    tlc07a

X10 1  2  4  5  3    tlc07a
R1  2  0   10K
R2  2  3   10K
R3  7  1   1K
C3  1  0   0.01u
R4  6  7   1K
C4  7  3   0.01u

*.PLOT AC (VP(3)-180) -360,90
.AC DEC 10 1000 1e+007
.TEMP 27
.PLOT AC VDB(1) VDB(2) VDB(3) -80,20
.PRINT AC VDB(1)  VDB(3)
.end
```

Figure x30111 Sallen-Key Low Pass Transfer Function

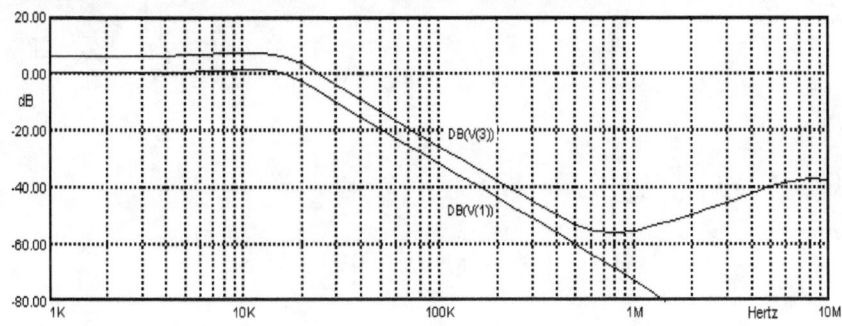

Experiment 4 Design an Op Amp High Pass Filter

The non-inverting op amp circuit relates V_3 to V_1.

(1) $\dfrac{V_3}{V_1} = 1 + \dfrac{R_2}{R_1} = K$

Straightforward circuit analysis produces the transfer function.

(5a) $\dfrac{V_3}{V_6} = \dfrac{Kp^2}{p^2 + (1/R_3C_3 + 1/R_3C_4 + (1-K)/R_4C_4)p + 1/R_3C_3R_4C_4}$

(5b) $\dfrac{V_3}{V_6} = \dfrac{Kp^2}{p^2 + (\omega_3 + \omega_{34} + (1-K)\omega_4)p + \omega_3\omega_4}$

(5c) $\dfrac{V_3}{V_6} = \dfrac{Kp^2}{p^2 + \dfrac{1}{Q}p + \omega_0{}^2}$

(6a) $\dfrac{1}{Q} = (\omega_3 + \omega_{34} + (1-K)\omega_4) \qquad \omega_0{}^2 = \omega_3\omega_4$

(6b) $R_4 = R_3 = 1K\Omega,\ C_4 = C_3 = 0.01\mu F \ \rightarrow\ \omega_3 = \omega_4 = \omega_0$

(6c) $\dfrac{1}{Q} = (2\omega_4 + (1-K)\omega_4) = \omega_0(3-K) \ \rightarrow\ K = 2 \ \rightarrow\ R_1 = R_2 = 10K$

(6d) $\dfrac{1}{Q} = \omega_0 \ \rightarrow\ $ (2c) $\dfrac{V_3}{V_6} = \dfrac{2p^2}{p^2 + \omega_0 p + \omega_0{}^2}$

A simplified Sallen-Key filter design process with a Butterworth response is as follows.

Figure X401 Sallen-Key High Pass Filter

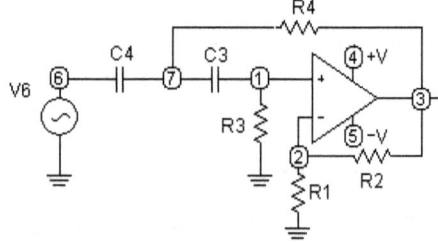

(7a) $C_4 = C_3 = 0.01\mu F$

(7b) $R_4 = R_3 = \dfrac{1}{2\pi f_0 C_3} = \dfrac{1}{2\pi \cdot (50/\pi) \cdot 10^3 \cdot 0.01 \cdot 10^{-6}} = 10^3 \Omega$

Build the filter, measure its performance, and write a Spice program to confirm performance.

Spice Program x4011
Figx4011.ckt Sallen-Key high pass Filter

```
VP 4 0 DC 5
VN 5 0 DC -5
V1 6 0 AC 1
.lib G:\!B\NLPbooks\ee109pdf\ecexspice\a_tlc07.txt

*.SUBCKT tlc07a
* Node assignments
*      |non-inverting input
*      |   |inverting input
*      |   |   |positive supply
*      |   |   |   |negative supply
*      |   |   |   |   output
*      |   |   |   |   |
*      1   2   4   5   3
*X1 0  2   4   5   3    tlc07a

X10 1  2   4   5   3    tlc07a
R1 2  0   10K
R2 2  3   10K
R3 1  0   1K
C3 7  1   0.01u
R4 7  3   1K
C4 6  7   0.01u

*.PLOT AC (VP(3)-180) -360,90
.AC DEC 10 10 1e+005
.TEMP 27
.PLOT AC VDB(1) VDB(2) VDB(3) -80,20
.PRINT AC VDB(1)  VDB(3)
.end
```

Figure x40111 Sallen-Key High Pass Transfer Function

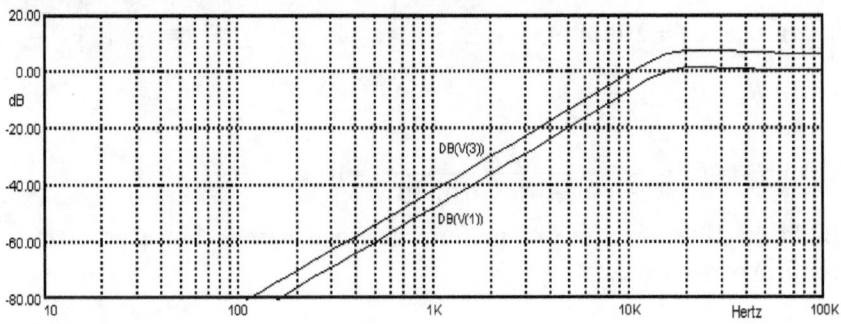

Experiment 5 Design an Op Amp Band Pass Filter

The non-inverting op amp circuit relates V_3 to V_1.

(1) $\quad \dfrac{V_3}{V_1} = 1 + \dfrac{R_2}{R_1} = K$

Straightforward circuit analysis produces the transfer function.

(8a) $\quad \dfrac{V_3}{V_6} = K \dfrac{1}{R_4 C_3} \times \dfrac{p}{p^2 + \left(\dfrac{1}{R_5 C_3} + \dfrac{C_3 + C_4}{R_4 C_3 C_4} + \dfrac{1}{R_3 C_4} + (1-K)\dfrac{1}{R_3 C_3} \right) p + \dfrac{R_3 + R_4}{R_5} \dfrac{1}{R_3 C_3 R_4 C_4}}$

(8b) $\quad \dfrac{V_3}{V_6} = K \omega_{43} \times \dfrac{p}{p^2 + (\omega_{53} + \omega_{434} + \omega_{34} + (1-K)\omega_3) p + \frac{R_3 + R_4}{R_5} \omega_3 \omega_4}$

(8c) $\quad \dfrac{V_3}{V_6} = K \omega_{43} \times \dfrac{p}{p^2 + \dfrac{1}{Q} p + \omega_0{}^2}$

(9a) $\quad \dfrac{1}{Q} = (\omega_{53} + \omega_{434} + \omega_{34} + (1-K)\omega_3) \qquad \omega_0{}^2 = \frac{R_3 + R_4}{R_5} \omega_3 \omega_4$

(9b) $\quad R_5 = 2K\Omega, \; R_4 = R_3 = 1K\Omega, \; C_4 = C_3 = 0.01\mu F \;\rightarrow\; 0.5\omega_{53} = 2\omega_{434} = \omega_{34} = \omega_3 = \omega_0$

(9c) $\quad \dfrac{1}{Q} = (3.5\omega_0 + (1-K)\omega_0) = \omega_0(4.5 - K) \;\rightarrow\; K = 4.6 \;\rightarrow\; R_1 = 10K \; R_2 = 36K$

(9d) $\quad \dfrac{1}{Q} = -0.1\omega_0 \;\rightarrow\; (2c) \; \dfrac{V_3}{V_6} = \dfrac{4.6\omega_0 p}{p^2 - 0.1\omega_0 p + \omega_0{}^2}$

A simplified Sallen-Key filter design process with a Butterworth response is as follows.

Figure X501 Sallen-Key Band Pass Filter

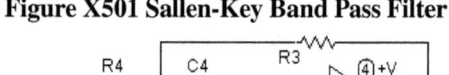
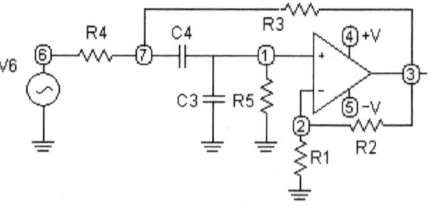

(10a) $\quad C_4 = C_3 = 0.01\mu F$

(10b) $\quad R_4 = R_3 = \dfrac{1}{2\pi f_0 C_3}$

(10c) $\quad R_4 = R_3 = \dfrac{1}{2\pi \cdot (50/\pi) \cdot 10^3 \cdot 0.01 \cdot 10^{-6}} = 10^3 \Omega$

Build the filter, measure its performance, and write a Spice program to confirm performance.

Spice Program x4011
Figx4011.ckt Sallen-Key Band pass Filter

```
VP  4  0  DC  5
VN  5  0  DC -5
V1  6  0  AC 1
.lib G:\!B\NLPbooks\ee109pdf\ecexspice\a_tlc07.txt

*.SUBCKT tlc07a
* Node assignments
*       |non-inverting input
*       |  |inverting input
*       |  |  |positive supply
*       |  |  |  |negative supply
*       |  |  |  |  output
*       |  |  |  |  |
*       1  2  4  5  3
*X1  0  2  4  5  3    tlc07a

X10  1  2  4  5  3    tlc07a
R1  2  0  10K
R2  2  3  36K
R3  7  3  1K
C3  1  0  0.01u
R4  6  7  1K
C4  7  1  0.01u
R5  1  0  2K

*.PLOT AC (VP(3)-180) -360,90
.AC DEC 200 100 1e+006
.TEMP 27
.PLOT AC VDB(1) VDB(2) VDB(3) -80,20
.PRINT AC VDB(1)  VDB(3)
.end
```

Figure x50111 Sallen-Key Band Pass Transfer Function

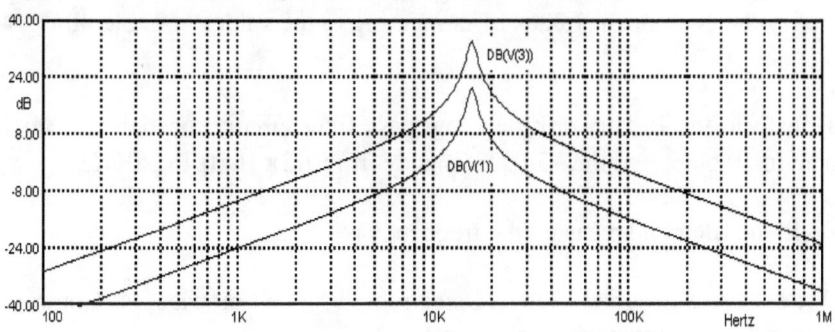

Experiment 6 Design a Butterworth Filter

1 Design an n=4 Butterworth low pass filter using the Z synthesis method. Verify that n =4 satisfies 1dB pass band loss, 30dB stop band loss, and a 3 to 1 stop band, −3dB pass band frequency ratio..

$$n \geq \frac{\log\left[\dfrac{10^{0.1\beta\,dB}-1}{10^{0.1\alpha\,dB}-1}\right]}{2\log(\omega_2/\omega_1)} = \frac{\log\left[\dfrac{10^{0.1\times 30\,dB}-1}{10^{0.1\times 1\,dB}-1}\right]}{2\log(3)} = \frac{\log\left[\dfrac{999}{0.2589}\right]}{2\log(3)} = \frac{\log 3858.25}{2\log 3} = 3.76$$

Verify the component values by comparing to the values in Table 401 n=4 on page 68.

Transform the Butterworth n=4 low pass filter circuit impedance from 1 ohm to 50 ohms and the −3dB frequency from 1/2π Hertz to 10KHz.

Build the filter and measure the frequency response.

Write a Spice program such as Spice program 4021 on page 64. Compare your measurements to the Spice results.

2 Convert the 10KHz low pass design to a high pass design.

Build the filter and measure the frequency response.

Write a Spice program such as Spice program 4021 on page 64. Compare your measurements to the Spice results.

3 Design an n=4 Butterworth low pass filter using the Darlington method. Verify the component values by comparing to the values in Table 404 n=4 on page 76.

Transform the Butterworth n=4 low pass filter circuit impedance from 1 ohm to 10 ohms and the −3dB frequency from 1/2π Hertz to 25KHz.

Build the filter and measure the frequency response.

Write a Spice program such as Spice program 4062 on page 74. Compare your measurements to the Spice results.

Experiment 7 Design a Bessel Filter

1 Design an n=4 Bessel low pass filter using the Z synthesis method. Verify the component values by comparing to the values in Table 403 n=4 on page 68.

Transform the Bessel n=4 low pass filter circuit impedance from 1 ohm to 50 ohms and the −3dB frequency from $1/2\pi$ Hertz to 5KHz.

Build the filter and measure the frequency response.

Write a Spice program such as Spice program 4032 on page 67. Compare your measurements to the Spice results.

2 Convert the 5KHz low pass design to a high pass design.

Build the filter and measure the frequency response.

Write a Spice program such as Spice program 4032 on page 67. Compare your measurements to the Spice results.

3 Design an n=4 Bessel low pass filter using the Darlington method. Verify the component values by comparing to the values in Table 406 n=4 on page 76.

Transform the Bessel n=4 low pass filter circuit impedance from 1 ohm to 50 ohms and the −3dB frequency from $1/2\pi$ Hertz to 20KHz.

Build the filter and measure the frequency response.

Write a Spice program such as Spice program 4062 on page 74. Compare your measurements to the Spice results.

Figure 301 Low Pass Filter Specification

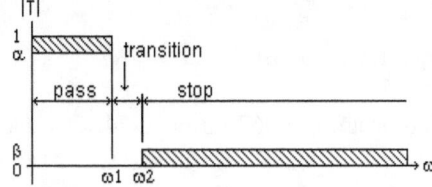

Experiment 8 Design a Chebyshev Filter

1 Design an n=5 Chebyshev low pass filter using the Z synthesis method. Verify that n =5 satisfies 1dB pass band loss, 40dB stop band loss, and a 1.8 to 1 stop band, −3dB pass band frequency ratio..

$$n \geq \frac{\cosh^{-1}\left[\dfrac{10^{0.1\beta\,dB}-1}{10^{0.1\alpha\,dB}-1}\right]^{\frac{1}{2}}}{\cosh^{-1}(\omega_2/\omega_1)} = \frac{\cosh^{-1}\left[\dfrac{10^{0.1\times40dB}-1}{10^{0.1\times1\,dB}-1}\right]^{\frac{1}{2}}}{\cosh^{-1}(1.8)} = \frac{\cosh^{-1}\left[\dfrac{9999}{0.2589}\right]^{\frac{1}{2}}}{1.1929} = 5.008$$

Verify the component values by comparing to the values in Table 402 n=5 on page 68.

Transform the Chebyshev n=5 low pass filter circuit impedance from 1 ohm to 50 ohms and the −3dB frequency from 1/2π Hertz to 8KHz.

Build the filter and measure the frequency response.

Write a Spice program such as Spice program 4031 on page 66. Compare your measurements to the Spice results.

2 Convert the 8KHz low pass design to a high pass design.

Build the filter and measure the frequency response.

Write a Spice program such as Spice program 4031 on page 66. Compare your measurements to the Spice results.

3 Design an n=5 Chebyshev low pass filter using the Darlington method. Verify the component values by comparing to the values in Table 405 n=5 on page 76. Transform the Chebyshev n=5 low pass filter circuit impedance from 1 ohm to 10 ohms and the −3dB frequency from 1/2π Hertz to 16KHz.

Build the filter and measure the frequency response.

Write a Spice program such as Spice program 4062 on page 74. Compare your measurements to the Spice results.

Index